Lecture Notes in Computer Science 4267

Commenced Publication in 1973
Founding and Former Series Editors:
Gerhard Goos, Juris Hartmanis, and Jan van

T0230032

Editorial Board

David Hutchison
 Lancaster University, UK
Takeo Kanade
 Carnegie Mellon University, Pittsburgh, PA, USA
Josef Kittler
 University of Surrey, Guildford, UK
Jon M. Kleinberg
 Cornell University, Ithaca, NY, USA
Friedemann Mattern
 ETH Zurich, Switzerland
John C. Mitchell
 Stanford University, CA, USA
Moni Naor
 Weizmann Institute of Science, Rehovot, Israel
Oscar Nierstrasz
 University of Bern, Switzerland
C. Pandu Rangan
 Indian Institute of Technology, Madras, India
Bernhard Steffen
 University of Dortmund, Germany
Madhu Sudan
 Massachusetts Institute of Technology, MA, USA
Demetri Terzopoulos
 University of California, Los Angeles, CA, USA
Doug Tygar
 University of California, Berkeley, CA, USA
Moshe Y. Vardi
 Rice University, Houston, TX, USA
Gerhard Weikum
 Max-Planck Institute of Computer Science, Saarbruecken, Germany

Ahmed Helmy   Brendan Jennings
Liam Murphy   Tom Pfeifer (Eds.)

Autonomic Management of Mobile Multimedia Services

9th IFIP/IEEE International Conference on Management
of Multimedia and Mobile Networks and Services, MMNS 2006
Dublin, Ireland, October 25-27, 2006
Proceedings

 Springer

Volume Editors

Ahmed Helmy
University of Florida
Department of Computer and Information Science and Engineering
Gainesville, FL 32611, USA
E-mail: helmy@ufl.edu

Brendan Jennings, Tom Pfeifer
Waterford Institute of Technology, Telecommunications Software & Systems Group
Cork Road, Waterford, Ireland
E-mail: bjennings@tssg.org, t.pfeifer@computer.org

Liam Murphy
University College Dublin, School of Computer Science and Informatics
Belfield, Dublin 4, Ireland
E-mail: liam.murphy@ucd.ie

Library of Congress Control Number: 2006934293

CR Subject Classification (1998): C.2, H.5.1, H.3, H.5, H.4

LNCS Sublibrary: SL 5 – Computer Communication Networks and
Telecommunications

ISSN   0302-9743
ISBN-10   3-540-47654-7 Springer Berlin Heidelberg New York
ISBN-13   978-3-540-47654-2 Springer Berlin Heidelberg New York

Springer is a part of Springer Science+Business Media

springer.com

© IFIP International Federation for Information Processing 2006
Printed in Germany

Typesetting: Camera-ready by author, data conversion by Scientific Publishing Services, Chennai, India
Printed on acid-free paper   SPIN: 11907381   06/3142   5 4 3 2 1 0

Preface

This volume presents the proceedings of the 9th IFIP/IEEE International Conference on *Management of Multimedia and Mobile Networks and Services (MMNS 2006)*, which was held from October 25th to 27th as part of Manweek 2006 in Dublin,Ireland. In line with its reputation as one of the pre-eminent fora for the discussion anddebate of advances in management of multimedia networks and services, the 2006 iteration of MMNS brought together an international audience of researchers and practitioners from both industry and academia.

One of the most significant trends of recent years has been the development and considerable market penetration of multimedia-capable mobile handsets. Nevertheless, significant research challenges remain in the area of management of mobile networks supporting multimedia *services*; challenges which must be addressed by the research community if the vision of ubiquitous availability of advanced multimedia services is to be realised. The MMNS Steering Committee, noting the growing interest within the research community on solving issues relating specifically to the transport of multimedia traffic over various mobile access technologies, made the significant decision to change the long name of the MMNS conference to reflect this change in focus. For2006 and subsequent years, it will be called "Management of Multimedia *and Mobile* Networks and Services".

In response to the broadening of the conference scope, the MMNS 2006 Technical Programme Committee issued a call for papers reflecting critical research issues in the area of management of multimedia and mobile networks. These issues included "traditional" MMNS topics such as: distributed multimedia service management, deployment of multimedia services, and protocols for multimedia services; together with new MMNS topics such as: seamless mobility of multimedia services, adaptive multimedia services, and management of wireless ad-hoc networks. Furthermore, the Technical Programme Committee decided that MMNS 2006 should have an overall theme of "Autonomic Management of Mobile Multimedia Services". This theme was chosen partially to reflect the broader Manweek 2006 theme of "Autonomic Component and System Management", but also due to the observation that one of the most promising avenues for advances in network and service management research is the application of concepts such as self-governance and self-organisation – concepts that are central to the autonomic management vision.

In response to the call for papers, a total of 71 paper submissions were received from the worldwide research community. Of these, 61 were full papers and 10 were short papers. A comprehensive review process was carried out by the Technical Programme Committee and additional subject area experts, with all papers receiving atleast three, and more commonly four, detailed reviews. Subsequently, all submissions were ranked based on review scores as well as the wider Technical Programme Committee's view of their contribution and relevance to the conference scope. After lengthy discussion, it was decided to accept 18 of the 61 submitted full papers (29.5% acceptance rate) and 6 short papers (3 of which were initially submitted as full papers, and 3that were selected from the 10 submitted short papers). These papers present novel and interesting contributions in topics ranging from video streaming to energy consumption models for ad-hoc networks, and from autonomic handover decision management to multimedia caching. We believe that, taken together, they provide a unique insight into the current state of the art in the management of multimedia and mobile networks and services.

There are many people whose hard work and commitment were essential to the success of MMNS 2006. Foremost amongst these are the researchers who submitted papers to the conference. The overall quality of submissions this year was high, and we regret that many high quality papers had to be rejected. We would like to express our gratitude to both the MMNS Steering Committee and the Technical Programme Committee, for their advice and support through all the stages of the conference preparation. We thank all paper reviewers, in particular those outside the Technical Programme Committee, for their uniformly thorough, fair and helpful reviews. We thank both IFIP and IEEE for their continued support and sponsorship of MMNS.

Most of the more time-consuming practical and logistical organisation tasks for the conference were handled by the members of the Manweek Organisation Committee – this made our jobs significantly easier, and for that we are very grateful. Finally, we wish to acknowledge the financial support of both Science Foundation Ireland and the Manweek corporate sponsors, whose contributions were hugely instrumental in helping us run what we hope was a stimulating, rewarding and, most importantly, an enjoy-able conference for all its participants.

October 2006

Ahmed Helmy
Brendan Jennings
Liam Murphy
MMNS 2006 TPC Co-chairs

Tom Pfeifer
Manweek 2006 Publication Chair

MMNS 2006 Organisation

Technical Programme Committee Co-chairs
Ahmed Helmy — University of Southern California, USA
Brendan Jennings — Waterford Institute of Technology, Ireland
Liam Murphy — University College Dublin, Ireland

Steering Committee
Nazim Agoulmine — University of Evry, France
Kevin Almeroth — University of California, Santa Barbara, USA
Ehab Al-Shaer — DePaul University, USA
Raouf Boutaba — University of Waterloo, Canada
Masum Z. Hasan — Cisco Systems, USA
David Hutchison — University of Lancaster, UK
Alan Marshall — Queens University of Belfast, UK
Guy Pujolle — Université de Pierre-et-Marie Curie, Paris 6, France
John Vicente — Intel, USA

Organisation Co-chairs
Brendan Jennings — Waterford Institute of Technology, Ireland
Sven van der Meer — Waterford Institute of Technology, Ireland

Publication Chair
Tom Pfeifer — Waterford Institute of Technology, Ireland

Publicity Co-chairs
Sasitharan Balasubramaniam — Waterford Institute of Technology, Ireland
John Murphy — University College Dublin, Ireland

Treasurer
Mícheál Ó Foghlú — Waterford Institute of Technology, Ireland

Local Arrangements
Miguel Ponce de León — Waterford Institute of Technology, Ireland
Dave Lewis — Trinity College Dublin, Ireland
Dirk Pesch — Cork Institute of Technology, Ireland
Gabriel-Miro Muntean — Dublin City University, Ireland
Seán Murphy — University College Dublin, Ireland
Rob Brennan — Ericsson, Ireland

Manweek 2006 General Co-chairs
William Donnelly — Waterford Institute of Technology, Ireland
John Strassner — Motorola Labs, USA

Manweek 2006 Advisors
Raouf Boutaba   University of Waterloo, Canada
Joan Serrat   Universitat Politècnica de Catalunya, Spain

MMNS 2006 Technical Programme Committee
Nazim Agoulmine   University of Evry, France
Kevin Almeroth   University of California, Santa Barbara, USA
Ehab Al-Shaer   DePaul University, USA
Pablo Arozarena   Telefónica Investigación y Desarrollo, Spain
Åke Arvidsson   Ericsson, Sweden
Chadi Assi   Concordia University, Canada
Fan Bai   General Motors Research Labs, USA
Sasitharan Balasubramaniam   Waterford Institute of Technology, Ireland
Javier Barria   Imperial College, UK
Bert-Jan van Beijnum   University of Twente, The Netherlands
Mohamed Bettaz   Philadelphia University, Jordan
Raouf Boutaba   University of Waterloo, Canada
Greg Brewster   DePaul University, USA
Alexander Clemm   Cisco Systems, USA
Nicola Cranley   Dublin Institute of Technology, Ireland
Spyros Denazis   University of Patras, Greece
Petre Dini   Cisco Systems, USA
Masum Z. Hasan   Cisco Systems, USA
Go Hasegawa   Osaka University, Japan
David Hutchison   University of Lancaster, UK
James Irvine   Strathclyde University, UK
Vana Kalogeraki   University of California, Riverside, USA
Ahmed Karmouch   University of Ottawa, Canada
Lukas Kencl   Intel Research Cambridge, UK
Georgios Kormentzas   University of the Aegean, Greece
Andrej Kos   University of Ljubljana, Slovenia
Jun Li   University of Oregon, USA
Hanan Lutfiyya   University of Western Ontario, Canada
Thomas Magedanz   Fraunhofer FOKUS, Germany
Alan Marshall   Queens University of Belfast, UK
Sven van der Meer   Waterford Institute of Technology, Ireland
Gabriel-Miro Muntean   Dublin City University, Ireland
Gerard Parr   University of Ulster, UK
George Pavlou   University of Surrey, UK
Andrew Perkis   Norwegian University of Science and Technology,
                                                              Norway
Dirk Pesch   Cork Institute of Technology, Ireland
Tom Pfeifer   Waterford Institute of Technology, Ireland
Guy Pujolle   Université de Pierre-et-Marie Curie, Paris 6, France
Reza Rejaie   University of Oregon, USA
Karim Seada   Nokia Research Labs, USA

Rolf Stadler	Royal Institute of Technology, Sweden
Burkhard Stiller	University of Zurich and ETH Zurich, Switzerland
John Strassner	Motorola Labs, USA
Marina Thottan	Bell Labs, USA
John Vicente	Intel, USA
Theodore Willke	Intel, USA
Lawrence Wong	National University of Singapore, Singapore
Theodore Zahariadis	Ellemedia Technologies, Greece
Roger Zimmerman	University of Southern California, USA
Michele Zorzi	University of Padova, Italy

MMNS 2006 Additional Paper Reviewers

Ibrahim Aloqily	University of Ottawa, Canada
Anatoly Andrianov	Motorola, USA
Steve Berl	Cisco Systems, USA
Ganesha Bhaskara	University of Southern California, USA
Dmitri Botvich	TSSG, Waterford Institute of Technology, Ireland
Björn Brynjúlfsson	Reykjavik University, Iceland
Lawrence Cheng	University College London, UK
Steven Davy	Waterford Institute of Technology, Ireland
Alan Davy	Waterford Institute of Technology, Ireland
Francesco De Pellegrini	CREATE-NET, Italy
Hajer Derbel	Université d'Evry, France
Gang Ding	Purdue University, USA
Debojyoti Dutta	University of Southern California, USA
Nikos Efthymiopoulos	University of Patras, Greece
Toby Ehrenkranz	University of Oregon, USA
Stenio Fernandes	CEFET-AL, Brazil
Joe Finney	Lancaster University, UK
William Fitzgerald	Waterford Institute of Technology, Ireland
Viktoria Fodor	KTH, Sweden
Vamsi Gondi	Université d'Evry Val d'Essonne, France
Alberto Gonzalez	Royal Institute of Technology, Sweden
Donna Griffin	Cork Institute of Technology, Ireland
Michael Guidero	University of Oregon, USA
Kamel Haddadou	Université de Pierre-et-Marie Curie, Paris 6, France
Evangelos Haleplidis	University of Patras, Greece
H. Harroud	University of Ottawa, Canada
Gavin Holland	HRL, USA
Wei-Jen Hsu	USC, USA
Dilip Krishnaswamy	Intel, USA
Elyes Lehtihet	Waterford Institute of Technology, Ireland
Cheng Lin	CMU, USA
Adam Lindsay	Lancaster University, UK

Table of Contents

Short Papers

Seamless Mobility

Bandwidth Provisioning and Control

Multimedia over Wireless (2)

Enhanced Multimedia and Real-Time Stream Throughput in Wireless Ad-Hoc LAN Using Distributed Virtual Channel Based MAC*

Md. Mustafizur Rahman and Choong Seon Hong**

Department of Computer Engineering, Kyung Hee University, South Korea
mustafiz@networking.khu.ac.kr, cshong@khu.ac.kr

Abstract. Recent and upcoming CSMA/CA based MAC protocols for wireless LANs offer block-acknowledgement or aggregated frame exchange in order to provide high data rate for multimedia and real-time data, while all of them avoid collisions by exclusive use of wireless medium that block neighboring nodes of both sender and receiver from participating in concurrent transmission; and thus downgrade medium utilization and overall throughput. In this work, we offer to create concurrent virtual channels over same physical channel by synchronizing the transmit/receive switching of senders and receivers and allow parallel data transfer. Each virtual channel is used by a transmitter/receiver pair and all virtual channels within 2-hop network utilize the medium by distributed coordination and avoid inter virtual channel interference. Simulation result shows that proposed scheme removes the obstruction to participate in parallel transmission for some neighboring nodes of sender and receiver and improves the network performance.

1 Introduction

Multimedia data and real time transmission necessitates increasing demand for reduced average latency at nodes and higher network throughput in wireless LAN environment. Wireless LANs use shared medium that requires an efficient channel access function for successful data transmissions. The choice of medium access scheme is difficult in ad-hoc networks due to the time varying network topology and the lack of centralized control [1]. The access problem when same receiver hears transmissions from many nodes has been much studied since the ALOHA, and it bounds on the throughput of successful collision-free transmissions and forces to formulate efficient transmission protocols [2]. Sharing channels in networks does lead to some new problems associated with hidden and exposed terminals. The protocol MACA [3] and its extension MACAW [4], use a series of handshake signals to resolve these problems to a certain extent. This has been standardized and used in DCF/EDCA/ADCF in IEEE 802.11 family of protocols. Such protocols prohibit all hidden and exposed nodes from participating in concurrent transmission (see fig. 1(a)) and hence packets

---

* This work was supported by MIC and ITRC Project.
** Corresponding author.

A. Helmy et al. (Eds.): MMNS 2006, LNCS 4267, pp. 1–12, 2006.

experience latency. Although this scheme supported the wireless LANs for more than a decade, it requires enhancement for increasing demand for higher network through-put (100mbps in 802.11n) for multimedia data. An important approach to improving transfer efficiency is provided with aggregate exchange (AE) sequences for the up-coming IEEE 802.11n that acknowledges multiple MPDUs with a single block ac-knowledgement (Block ACK) in response to a block acknowledgement request (BAR)[5][6]. This protocol effectively eliminates the contention for every MPDU. If attempting to use the existing MAC protocols without aggregation, a PHY rate of 500 Mbps would be required to achieve the throughput goal of 100 Mbps at the MAC Service Access Point (MAC SAP) [7]. The aggregated exchange (or the block ACK in 802.11e) with conventional RTS/CTS mechanism, however, increases latency for data/multimedia packets at neighboring nodes, because of prolonged delay before channel access.

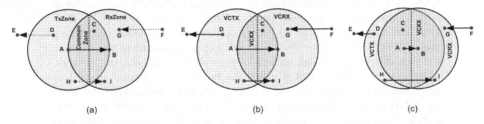

(a)   (b)   (c)

Fig. 1. (a) Conventional Channel reservation where transfer between A and B is possible only; (b) Proposed Virtual Channel Based transfer allowing multiple concurrent transmission; and (c) Virtual Channel scheme when A and B is closer to each other than (b)

In this work, we try to enhance the 802 family of MAC protocols toward distrib-uted spatial reuse of wireless medium for the aggregated exchange or block ACK schemes so that multimedia and real-time data can be transferred in parallel. Our pro-posed protocol creates separate virtual channel (VC) for every sender-receiver pair and multiple VCs can coexist within same physical channel in 2-hop networks; and, packets can be transferred over these virtual channels concurrently reducing the aver-age latency of the network and increased network throughput.

This rest of this paper is organized as: related works has been described in section 2; section 3 introduces the virtual channel concept and describes how they can coexist without interfering each-other, section 4 describes the mechanism for VC-MAC. Sec-tion 5 gives an analytic model for VC delay overheads and discrete system simulation and numerical results obtained from the simulation are presented and described in section 6. Finally, section 7 concludes the paper and contains scope for future work.

2   Related Works

Numerous works supporting concurrent transmissions have been done to improve the capacity in 2-hop network. Sigh et al. [8] proposed a distributed spatial reuse (DSR) MAC protocol for IEEE 802.11 ad hoc wireless LANs to increase bandwidth utiliza-tion and reduce power consumption using power controlled transmission. The power controlled schemes, however, depend on enhancement in the physical layer. In addition to this, finding the maximum independent set (MIS) from the Interference Graph (IG), obtained from IIM signals, is an NP-complete problem and it adds complexity at the MAC layer.

Acharya et al. [9] presents the initial design and performance study of MACA-P, a RTS/CTS based MAC protocol, that enables simultaneous transmissions in multi-hop ad-hoc wireless networks. Like 802.11, MACA-P uses a contention-based reservation phase prior to transmissions. The data transmission is delayed by a control interval, which allows multiple sender-receiver pairs to synchronize their data transfers, thereby avoiding collisions. The enhanced version of MACA-P [10] that they have devised with adaptive learning, avoids the collisions caused by MACA-P's attempts at concurrent transmissions. The control gap in the proposed MACA-P protocol uses for every packet transfer, and hence it increases the overhead.

Rozovsky et al. [11] proposed another protocol for media access control in ad hoc networks that seeks to avoid collisions without making explicit reservations for each and every packet. They used a random schedule driven by a pseudo-random number generator and the nodes publish their schedules to all hidden as well as exposed nodes. This allows each node to opportunistically choose transmission slots. But the protocol would work better for the hardware where synchronized slot is available.

Directional antenna based several solutions have been proposed [12, 13]. However, the hardware cost for using directional antennas to achieve spatial reuse is too high.

Our scheme is proposed as an enhancement of the MAC layer protocols that can be implemented over existing physical layers with better throughput for multimedia and real-time data streams. The complexity in synchronization between nodes is small compared to other proposed protocols.

3 The VC-MAC

The major confronting element in supporting parallel use of the channel in the wireless MAC protocols is to avoid interference at the receiver. We offer transferring data over half-duplex virtual channels between each sender-receiver pair, to support parallel transfer. Before sending data, the sender and the receiver (hereinafter referred as Originator and Recipient) establish the virtual channel and then transfers data. Concurrent virtual channels within 2-hop is allowed when they are synchronized with each other. Each virtual channel is closed after completion of data transfer.

3.1 Virtual Channel Concept

A virtual channel is a series of alternating Data and Control windows (DW and CW) providing a half-duplex connection between sender and receiver in order to transfer blocks of data packets (referred as data frames) using aggregated frame or block ACK

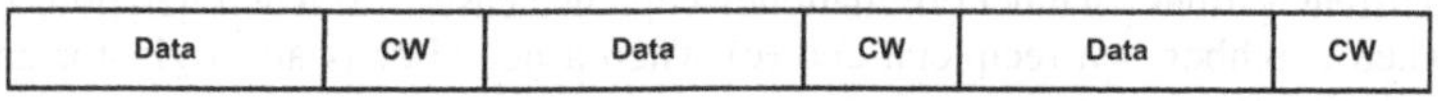

Data	CW	Data	CW	Data	CW

Fig. 2. The construct of a virtual channel

scheme. The originator transmits data frames one after one during the data window; and the recipient acknowledges received packets during the control windows. Synchronized concurrent virtual channels in neighboring nodes allow non-interfering data transfers within 2-hop network.

3.2 Concurrent Virtual Channels

In order to introduce spatial reuse of wireless medium, we allow concurrent virtual channels in several cases. To determine whether two nodes can establish a virtual channel, we introduce a new variable *vcState* at each node. The *vcState* indicates the existence and type of virtual channels in neighboring nodes. The *vcState* values and corresponding meaning are given in the following table:

Table 1. Values of *vcState* Variable

Value	Description
VCNO	No VC exist within node's neighborhood
VCTX	One or more in-sync (with common control window) VC Originator(s) exist within its neighborhood
VCRX	One or more in-sync (with common control window) VC Recipient exist within its neighborhood
VCXX	More than one of different type (RX/TX) or out of sync Virtual Channels exist within neighborhood

In-Sync VCs: If more than one same type of VC participants (originator or recipient) exist within the neighborhood of a node, where the data windows in each VC start simultaneously, then all such VCs are said to be in-sync with each other. Trivially, single VC within neighborhood of any node is said to be in-sync. In the following figure 3, neighbors O_1 and O_3 of node O_x are originators of two VCs. If the data and control windows of these two VCs start simultaneously, then these two VCs will be in-sync VCs.

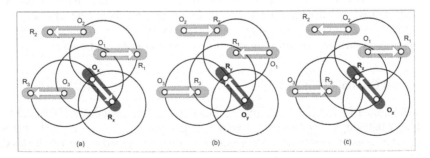

Fig. 3. Concurrent Virtual Channel origination/acceptance cases: (a) when neighbor is an originator, (b) when neighbor is a recipient, and (c) when a neighbor is an originator and another neighbor is a recipient

When a node finds its *vcState* value either VCNO or VCTX and it has data stream to send, it can originate the VC creation process; and the recipient can accept a VC formation request depending on its *vcState* value. Nodes having VCNO value can accept any request whereas VCRX valued nodes can accept VC request if the new VC can be in-sync with the other VC at recipient's neighborhood. In Fig. 3(a) and 3(b) O_1 and O_3 originated VCs because they didn't have any neighboring node participating in any other VC. O_2 finds a neighbor (with VCTX state) with one VC in fig 3(a) so it can originate and aligned (in-sync) with that virtual channel whereas O_x in the same figure finds two VCs within its neighborhood and may be with different data and control window timing (out of sync), so it can have either VCXX or VCTX (barely). In case of VCXX, O_x cannot originate a VC establishment. But if the order of origination is reverse (O_x creates VC earlier than O_3) then both of them would have VCTX in *vcState* and concurrent virtual channels could be established. Similarly, if the *vcState* value at R_y in figure 3(b) is VCRX then it can accept the VC establishment request from O_y. In the figure 3(c), the R_z node has two neighbors with VCs but one is originator and the other is recipient. So, it is impossible to originate or accept request for new VC establishment. The *vcState* value for such nodes becomes VCXX when VC participations found heterogeneous within neighborhood. This is also true for common neighbors of the originator and recipient of same VC.

The formation of virtual channels is completely distributed; local situations in each 2-hop network would allow/deny the concurrent virtual channels. The local VC creation order takes vital role in VC chaining.

4 The Mechanism and Operational Strategies

As mentioned earlier, a VC is a point to point half-duplex channel where the originator and recipient synchronize themselves by predefined time periods. The entire data transfer over VCs is a 3-phase process: Virtual Channel Opening, Transfer, and Closing down. Five new control frames (VCREQ, VCRES, VCCFM, VCDEL, and VCDCK response) are offered to handle the virtual channel establishment and closing phases. Data transfer can be done using existing BAReq and BA frames in IEEE 802.11e MAC protocol or aggregated frame scheme offered for proposed 802.11n.

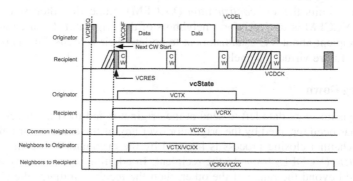

Fig. 4. Virtual Channel Signals and vcState values

4.1 Virtual Channel Opening

The originator, upon obtaining the transmission opportunity (TXOP), initiates the virtual channel opening phase by sending a *VCREQ* (Vritual Channel Request) frame to the intended receiver (recipient). The *VCREQ* frame contains data stream size, proposed Data Window size and *vcState* of the originator. According to the conditions described in section 3.2, the recipient determines whether it is able to accept the request or not. The recipient replies with *VCRES* (VC Request Response) only when an in-sync virtual channel formation is possible. The *VCRES* frame contains data window size, next control window start time and receive-buffer size. Recipients having VCNO state starts the control window immediately, and the *VCRES* frame is sent to the originator in the control window. On the other hand, the recipients with VCRX state wait for next control window for other VC(s) within its neighborhood and sends *VCRES*. The *VCRES* frame from VCRX recipients contains modified data-window size, next control-window time to make the VC in-sync with other VCs. Neighbors of the recipient update their *vcState* value with VCRX upon receipt of the VCRES frame; and common neighbors of both originator and recipient turn into VCXX state.

The originator then waits for the next data window and then sends the VC confirm frame (VCCFM) followed by first block of the data stream. The VCCFM is used to inform neighbors of the originator about new virtual channel. Neighbors of the originator update the *vcState* value to VCTX or VCXX and can participate in data transfer by forming another synchronized virtual channel only.

4.2 Data Transfer

After virtual channel is established, the originator sends data stream in a series of blocks separated by control windows during the data windows. Each burst of data block contains several data frames interleaved by shortest inter frame space period (SIFS in IEEE 802.11). Recent developments in physical layer inspired researchers to adapt block acknowledgement in MAC layer and recently IEEE Computer Society released the IEEE 802.11e MAC protocol supporting block acknowledgement [1]. In VC-MAC we follow the immediate policy of block acknowledgement scheme in the 802.11e standard.

Before sending the first data frame in the first data window of a virtual channel, the originator transmits the *VC confirmation* (VCCFM) frame first then starts the data burst. This VCCFM is sent to inform neighbors of the originator that are out of range from the recipient about VC establishment to update their *vcState*. This is required to synchronize future virtual channels.

4.3 Closing Down

When there is no more data left to send at the originator and the final block acknowledgement has been received by the originator over any of the existing virtual channel; the virtual channel closing process is initiated by the originator by sending a *virtual channel delete* (VCDEL) frame to the recipient. In addition to this, if any of the stations moved beyond the range of the other, then the receiver initiates the closing operation. The VCDEL is acknowledged by VCDCK by the recipient during the next control window.

4.4 Joining EDCA

After closing a virtual channel down, the originator or recipient may contend to get access to the medium and start sending data packets in presence of other virtual channels resulting in error in received data at neighboring recipients with virtual channels. In order to eliminate such situations, contending the medium for data transmission after closing virtual channel is delayed for next data/control windows and the delaying continues until it receives VCDEL or VCDCK frame from other originator or receiver.

In such cases, each participant in virtual channels or the neighbors of such participants takes different actions. The originator senses the medium just after issuing the VCDEL and if it finds any transmission, its virtual channel is closed but it continues the VCTX state. However, if it does not receive any signal, it turns into the VCNO state. This is same for other neighbors in the originator side only. In the recipient side, as soon as the recipient gets the control window, it closes the virtual channel but continues the VCRX state for next control windows until it receives the VCDCK frame from the last recipient. If it gets any corrupt data during next control windows; it is inferred that there exists other virtual channels within its neighborhood and it continues with VCRX state. Neighbors of both originator and recipient turns into VCNO state after it receives the VCDSK from last recipient (if any).

```
Procedure JoinEDCA(var vcState)
Var data;
Begin
   If( vcState = VCTX)
       while not DataWindow do no-op;
       data := ReceiveFrame();
       if data = notCorrupt and
         (data=noReceivedData or data=VCDEL) then
          vcState := VCNO;
       else
          Call JoinEDCA (vcState);
   Else if (vcState = VCRX)
       while not ControlWindow do no-op;
       data := ReceiveFrame();
       if data = notCorrupt and
         (data=noReceivedData or data=VCDCK) then
          vcState := VCNO;
       else
          Call JoinEDCA (vcState);
   Else if (vcState = VCXX)
       data := ReceiveFrame();
       if data = notCorrupt and
         (data=noReceivedData or data=VCDEL) then
          vcState := VCRX;
          Call JoinEDCA(vcState);
       Else if data = notCorrupt and data=VCDCK) then
          vcState := VCTX;
          Call JoinEDCA(vcState);
   End;
```

Fig. 5. Algorithm for the receivers to join the EDCA contention after closing the virtual channel

5 Analysis

5.1 Assumptions

We consider a 2-hop wireless ad-hoc network having stations negligible propagation delay. Influence of packet retransmissions is out of scope for this work as we select errorless packet delivery by the physical layer. In lossy channels, the analysis becomes more complex which should be investigated in the future work. The multimedia stream (data) generation pattern follows the Poisson distribution functions at any station and stream (data) sizes are uniformly distributed.

5.2 Overhead Modeling

Channel Establishment Cost: Time required for the setup phase for a VC is:

$$T_{VCsetup} = T_{VCREQ} + t_{remainDW} + T_{CW} \tag{1}$$

where, T_{VCREQ} is the time required to transmit *VCREQ* Frame, $T_{remainDW}$ is the remaining time of the current Data Window after receiving the *VCREQ*, T_{CW} is the Control Window period.

The $T_{remainDW}$ is random and depends on the arrival time of the *VCREQ* frame in current data window. The value of $T_{remainDW}$ is between 0 and data window period. Hence, the total overhead for the setup phase compared to 802.11e Block Acknowledgement scheme is:

$$\tau_{setup} \approx t_{remainDW} + T_{CW} \tag{2}$$

Data Transfer Costs: For a data stream requiring n data bursts consisting of s_i ($i | i \in 1 \cdots n$) data frames in the i-th data burst, the required time in slots over a virtual channel is-

$$T_{tr} = \sum_{i=1}^{n} \left(s_i \times T_{Data} + t_{remainDWi} + T_{CW} \right) \tag{3}$$

where, $t_{remainDWi}$ is the remaining time in the data window for the i-th data-burst, and T_{Data} is the average data-frame transmit time.

From the equation for transferring data frames, overhead in time for the latency experienced due to delayed responses from the receiver compared to the immediate block-acknowledgement mode becomes:

$$\tau_{transfer} = \sum_{i=1}^{n} \left(T_{remainDW_i} + T_{CW} \right) \tag{4}$$

For simplicity, we consider the average remaining time in each data window as $T_{remainDW}$, and the total overhead for transferring data of single data stream over each virtual channel becomes,

$$\tau_{transfer} = nT_{remainDW} + nT_{CW} \tag{5}$$

Therefore, the overhead can be reduced if and only if we can limit the control window size as minimum as possible and if each data stream can maximize the utilization of the data window. But the control window size cannot be minimized below the time slots required to send Block Acknowledgements with highest frame size.

For m concurrent virtual channels sharing common Control Window; the average overhead becomes

$$\tau_{transfer} = \frac{n}{m}(T_{remainDW_i} + T_{CW}) \tag{6}$$

Channel Closing Cost: Required time to close a virtual channel is calculated from:

$$T_{clo\sin g} = T_{VCDEL} + T_{remainDW} + T_{CW} \tag{7}$$

where, T_{VCDEL} is the transmitting time for the VCDEL request.

The total time required to close the virtual channel and join the EDCA contention is:

$$T_{Join} = T_{clo\sin g} + (n-1)(T_{DW} + T_{CW}) \tag{8}$$

where n is the number of concurrent virtual channels before closing the channel. Therefore, the overhead due to closing of the virtual channel is:

$$\tau_{join} = T_{remainDW} + T_{CW} + m \times (n-1)(T_{DW} + T_{CW}) \tag{9}$$

Total time required to send n data streams over n concurrent virtual channel is given by,

$$T_n^{vc} = n \times T_{setup} + T_{transfer} + T_{join} \tag{10}$$

6 Simulation

The VC-MAC protocol has been simulated on x, y grid topology; from 2×2 to 5×6 grids and compared with recent 802.11e block acknowledgement (BA) scheme. The channel is assumed to be reliable enough to transfer data without retransmission as discussed in 5.1. Our simulation considers parameters as follows. The data rate at physical layer is 54Mbps and in the MAC layer IEE802.11e EDCA and Block Acknowledgement (BA) scheme is used without prioritization. Data-frame size was fixed to 5K bytes. The entire DW-CW cycle in a virtual channel is 3000μs, where 2965μs allocated to data window and the rest 35μs for control window. The control window duration is sufficient to transmit VCRES, BA and VCDCK signals. Multimedia data stream sizes vary from 5K bytes to 250 K bytes and their arrivals at nodes follow Poisson distribution.

The rate of added latency for each virtual channel due to delay in initial response and the control window intervals compared to block acknowledgement BA scheme in 802.11e is depicted in Fig. 6(a). Simulation result shows that the added latency (mean delay) in each virtual channel decreases as the multimedia stream size increases for fixed data window duration. Latency in VCs reduces with increased stream size because of smaller influence of $t_{remainDW}$ over total turnaround time. For streams of 100K bytes, the added latency is approximately 10% and less than 30% for the first and consecutive VCs in each 2-hop network respectively. Corresponding reduction in data rate at each virtual channel is presented in the Fig 6(b). Virtual channel capacity increases with increased stream size, and approaches to the physical channel capacity. The patterns in both Fig. 6(a) and 6(b) follow the analytical approach as described in section 5.2.

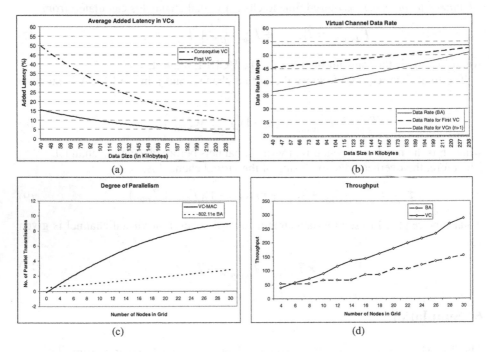

Fig. 6. Simulation Results: (a) Added latency in VCs, (b) Data rate in VCs, (c) Observed degree of parallelism in Grids, and, (d) Network throughput

The curve in Fig. 6(c) shows the average degree of parallelism obtained for the grid networks. The number of parallel VCs increases at higher rate than that of IEEE 802.11e BA scheme with increased number of nodes in the grid. Fig. 6(d) shows the influence in throughput for the degree of parallelism obtained. Control window durations in VCs reduces the network throughput due to reduction in channel capacity, however, overall throughput of the network increases when the network can create concurrent virtual channels within 2-hops. The proposed VC-MAC offers better throughput because it allows hidden and exposed terminals to participate in parallel transmissions, whereas conventional RTS/CTS bases MAC protocols cannot.

7 Conclusion

This work investigates the bottleneck in improving medium utilization for multimedia streams in WLAN environment. It is identified that conventional CSMA/CA MAC, that uses RTS/CTS handshaking, forbids all hidden and exposed nodes to participate in parallel transmissions. Denial of such parallel transmissions in these protocols, imposed due to transmit/receive switching of the nodes during data transmissions, increases latency for multimedia streams. Our investigation identifies several cases where parallel transmissions of multimedia streams are possible; and proposes the virtual-channel based MAC (VC-MAC) to decrease the packet latency and improve overall network performance. The VC is a schedule of synchronized transmit/receive switching of the participating nodes that allows hidden or exposed nodes to initiate parallel transmissions. The synchronization technique avoids collisions at receiving nodes.

Proposed VC-MAC scheme is analyzed and simulated on grid topology, and then results are compared with the Block Acknowledgement scheme of IEEE 802.11e standard for large data or multimedia streams. Analysis and simulation shows that VC-MAC can successfully avoid interference at the receivers and can allow hidden and exposed terminals to create parallel virtual channels. This scheme improves the overall network throughput although it reduces the data rate of each connection pair (virtual channel).

In this work, the evaluation of VC-MAC is done in an error-free environment. The value or otherwise of a MAC tends to show when it is subjected to errors. Impact of retransmissions on network capacity or throughput and finding the optimum data window size and incorporating service differentiation for QoS are open issues for future work.

References

[1] S. Ramanathan, and M. Steenstrup. A survey of routing techniques for mobile communication networks. Mobile Networks and Applications, Vol. 1, no 2, pp. 89-104, 1996.
[2] R. Gallager. A perspective on multi-access channels. IEEE Trans. Infrom. Theory (Special issue on Random Access Communications), Vol. IT-31, Mar. 1985.
[3] P. Karn. MACA: A new channel access method for packet radio. In proceedings of the 9th Computer Networking Conference, pp. 134-140, 1990.
[4] V. Bharghavan, A. Demers, S. Shenker, and L. Zhang. MACAW: A media access protocol for wireless LANs. In Proceedings of the ACM SIGCOMM'94, pages 212–215, London, UK, Aug. 1994.
[5] Wireless LAN Medium Access Control (MAC) and Physical Layer (PHY) specifications (Part 11), Amendment 8: Medium Access Control (MAC) Quality of Service Enhancements, Std. 802.11e, Nov. 2005.
[6] Data Link Frame Aggregation Protocol, J. Michael Meehan, Philip A. Nelson, David Engeset and Athos Pashiardis, Information: An International Interdisciplinary Journal, Volume 7 Number 1, pp. 59-68, January 2004.
[7] J. M. Wilson. The Next Generation of Wireless LAN Emerges with 802.11n, Intel Magazine, URL http://www.deviceforge.com/articles/AT5096801417.html, 2003.

[8] K. P. Sigh, C. M. Chou, M. Y. Lu, and, S. M. Chen. A Distributed Spatial Reuse (DSR) MAC Protocol for IEEE 802.11 Ad-Hoc Wireless LANs. In the proceedings of the 10th IEEE symposium on Computers and Communications, ISCC 2005, pages 658-663, 2005.

[9] A. Acharya, A. Misra, and S. Bansal. MACA-P: A MAC for concurrent transmissions in multi-hop wireless network. In proceedings of the IEEE International Conference on Pervasive Computing and Communications, pages 505-508, 2003.

[10] A. Acharya, A. Misra, and S. Bansal. Design and Analysis of a Cooperative Medium Access Scheme for Wireless Mesh Networks. In proceedings of the First International Conference on Broadband Networks (BROADNETS'04), pages 321-631, 2004.

[11] R. Rozovsky, and P. R. Kumar. SEEDEX: A MAC protocol for ad hoc networks. In proceedings of ACM Mobihoc, 2001.

[12] R. R. Chowdhury, X. Yang, N. H. Vaidya, and, R. Ramanathan. Using directional antennas for medium access control in ad hoc networks. In Proceedings of the ACM international Conference on Mobile Computing and Networking, pages 59-70, 2002.

[13] Y. Wang, and J. J. Garcia-Luna-Aceves. Spatial reuse and collision avoidance in ad hoc networks with directional antennas. In proceedings of the IEEE Global Telecommunications Conference, volume 1, pages 112-116, 2002.

Adaptive Streaming of Combined Audio/Video Content over Wireless Networks

Jeffrey Kang, Marek Burza, and Peter van der Stok

Philips Research, High Tech Campus 31, 5656 AE Eindhoven, The Netherlands
{Jeffrey.Kang, Marek.Burza, Peter.van.der.Stok}@philips.com

Abstract. This paper describes a method for robust streaming of combined MPEG-2 audio/video content over in-home wireless networks. We make use of currently used content distribution formats and network protocols. The transmitted bit-rate is constantly adapted to the available network bandwidth, such that audio and video artifacts caused by packet loss are avoided. Bit-rate adaptation is achieved by using a packet scheduling technique called I-Frame Delay (IFD), which performs priority-based frame dropping upon insufficient bandwidth. We show an implementation using RTP and an implementation using TCP. Measurements on a real-life demonstrator set-up demonstrate the effectiveness of our approach.

1 Introduction

In-home networks are becoming more and more common in consumer households. They connect together the different electronic devices in the home from the CE, mobile and PC domains, with which different digital content is stored and viewed. Such in-home networks typically consist of wired and wireless segments. Especially wireless networks are popular due to the ease of installation. However, wireless networks cannot always deliver the bandwidth needed for transporting the desired content. This is because wireless networks offer a lower bandwidth than wired networks, and very often this bandwidth has to be shared between multiple streams. Furthermore, the wireless network bandwidth often exhibits an unpredictable and fluctuating behavior. This is mainly caused by reflections and obstacles, roaming portable devices, and disturbances from other devices (e.g. neighboring wireless networks, microwaves). Other network technologies, e.g. power-line, also exhibit similar stochastic behavior under interference. For streams with real-time requirements, such as audio/video (A/V), insufficient network bandwidth causes data loss and the associated artifacts, such as blocky distortions in the image and disturbances in the audio. This is clearly unacceptable for the end user.

The problem of bandwidth sharing can be addressed by several QoS (Quality-of-Service) techniques such as prioritization and bandwidth reservation and access control mechanisms. In this paper, we address the second problem, namely that of fluctuating bandwidth for audio/video streaming. A technique is presented to stream the A/V content over the network, whereby the amount of data transmitted by the sender is dynamically adapted to the available bandwidth by selectively dropping data. In this way the perceived quality of the A/V stream is dynamically adjusted according to the quality of the network link. Our solution can be implemented using RTP (Real-time Transport Protocol) and TCP (Transmission Control Protocol).

A. Helmy et al. (Eds.): MMNS 2006, LNCS 4267, pp. 13–24, 2006.
© IFIP International Federation for Information Processing 2006

This paper is organized as follows. Section 2 lists some related work. Section 3 describes our adaptive streaming technique and its implementations. Experimental results are presented in Section 4. This paper ends with conclusions and final remarks.

2   Related Work

Most work in literature concentrates around streaming of video only, whereas we also consider audio, which is streamed together with the video stream. Most users are more sensitive to audio than to video artifacts, therefore these should be avoided. Different approaches have been proposed for adaptively streaming video over wireless networks. These solutions can be categorized by:

- *Scalable video*. The original video is encoded in different layers, where a *base layer* contains video of acceptable quality, and one or more *enhancement layers* enhance the quality of the base layer. Adaptation is done by dropping enhancement layers in case of insufficient bandwidth. As long as the base layer can be transmitted without loss, no artifacts will occur. Examples of this approach can be found in [1][2][3].
- *Transcoding/transrating*. Here the original bit-rate of the video is adaptively changed to a different bit-rate depending on the available bandwidth, e.g. by dynamically changing the quantization parameters. Examples of such solutions are found in [4][5].
- *Frame dropping*. Complete frames are dropped in case of insufficient bandwidth, examples are found in [6][7].

Our adaptation technique falls into the last category and is called *I-Frame Delay* (IFD), previously reported in [8]. This paper extends this by including audio. IFD is a scheduling technique which drops MPEG video frames when detecting insufficient bandwidth, thereby favoring I- and P-frames over B-frames. Artifacts are avoided as long as only B-frames are dropped. The result is a decreased video frame rate. The perceived quality here is lower than e.g. scalable video [9] or transrating, however IFD has the lowest cost in the implementation, and is able to react quickly to the changing conditions. Most MPEG decoder implementations are able to decode the resulting stream. Furthermore, the decision to perform adaptation is based on send buffer fullness, instead of relying on feedback from the receiver for bandwidth estimation. Therefore no receiver modifications are necessary. The work of [10] resembles ours the most, dealing with adaptive streaming of MPEG-2 Transport Streams by means of frame dropping. However, this approach relies on specific receiver feedback about the network status. Furthermore, since the feedback arrives periodically, the beginning of an abrupt bandwidth drop can often not be handled in time, leading to a short burst of lost packets (and hence frames). Lastly, only RTP streaming is considered, whereas we also consider TCP.

3   Adaptive Streaming of MPEG-2 Transport Streams

In this paper, we focus on the home scenario where content is entering the home from a content provider via ADSL, cable or satellite, and possibly stored on a home storage server. Here the most common content format is MPEG-2 Transport Streams (TS) [11]. Therefore we will further discuss adaptive streaming of MPEG-2 TS. Section 3.1 describes our IFD adaptive video scheduling technique. Sections 3.2 and 3.3 explain how this technique can be applied on Transport Streams using RTP and TCP.

3.1 I-Frame Delay

I-Frame Delay (IFD) is a scheduling technique for adaptive streaming of MPEG video. The basic idea is that scheduler will drop video frames when the transmission buffer is full and overflow is imminent due to insufficient bandwidth, to reduce the transmitted bit-rate. The algorithm is characterized by the following: 1) buffer fullness is indicated by the *number of frames* currently in the buffer (not the number of bytes), 2) less important frames (B-frames) are dropped in favor of more important frames (I- and P-frames), 3) the transmission of I-frames is delayed when conditions are bad but as little as possible omitted even if they are out-of-date w.r.t. the display time, because they can still be used to decode subsequent inter-predicted frames[1].

The IFD mechanism is based on two parts: 1) during parsing and packetization of the stream into network packets, the stream is analyzed and the packets are *tagged* with a priority number reflecting the frame type (I/P/B)[2], and 2) during transmission, packets are *dropped* by the IFD scheduler when the bandwidth is insufficient. Non-video packets are given a priority number not recognizable by the IFD scheduler, which therefore will never drop them. The size of the IFD buffer should be big enough to hold a number of frames. The minimum required size is two frames, one to hold the frame currently being sent (denoted as S), and one currently waiting to be sent (denoted as W). Increasing the IFD buffer size permits more intelligent decisions on which frames to be dropped, at the cost of increased latency and memory usage. Figure 1 depicts an example of the buffer filling. The numbers represent the priority numbers of the packets. In this example the IFD-related priority numbers are 10 and higher. The packets with priority number 12 belong to the S frame, and the packets with number 11 belong to the W frame. Currently a packet from an incoming frame C is about to enter the buffer. As can be seen, it is possible to interleave video packets with non-video packets (priority numbers 0 and 2). When a packet belonging to C tries to enter the IFD buffer and both

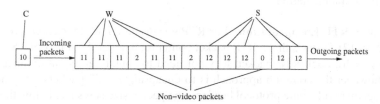

Fig. 1. Network packets in the IFD buffer

---

[1] We assume that out-of-date frames are still decoded and only thrown away by the renderer, rather than thrown away already before decoding.

[2] Since there may be multiple consecutive B-frames between two P-frames, we use different tags to distinguish between them.

S and W are full, the IFD scheduler will decide to drop a frame (in C or W), based on the priority numbers. No artifacts will occur if only B-frames are dropped, because no subsequent frames depend on them. When the network conditions are such that also P-frames are dropped, then the GOP (Group of Pictures) is said to be *disturbed* and the remainder of the GOP is also dropped. This causes the effect of the image being temporarily frozen, the duration of which depends on the GOP size. For an IFD buffer capable of holding two frames, the frame dropping algorithm is shown below.

```
 1: procedure BUFFER_ENQUEUE(C)
 2:     if Disturbed_GOP == True then
 3:         if C is of type I then                          ▷ New GOP is encountered
 4:             Disturbed_GOP = False                       ▷ Reset disturbed GOP flag
 5:         end if
 6:     end if
 7:     if Disturbed_GOP == True then
 8:         Discard C                                       ▷ Discard rest of disturbed GOP
 9:         return
10:     end if
11:     if W is empty then
12:         Store C in W
13:     else
14:         if C is of type I then
15:             Overwrite W with C
16:         else if C is of type B then
17:             Discard C
18:         else if W is of type I or P then
19:             Discard C
20:             if C was of type P then                     ▷ Discarded frame is P-frame
21:                 Disturbed_GOP = True                     ▷ Set disturbed GOP flag
22:             end if
23:         else
24:             Overwrite W with C
25:         end if
26:     end if
27: end procedure
```

3.2 RTP Implementation

This section describes our adaptive streaming approach using RTP. We first present a method for encapsulating TS packets into RTP packets, such that they can be used by IFD. This does not come without consequences, as will be explained. We then describe our streamer implementation.

IFD-Friendly RTP Encapsulation. For RTP encapsulation of TS packets we adhere to RFC2250 [12]. It states that a RTP packet contains an integral number of TS packets (188 bytes each). Each RTP packet does not need to contain an equal number of TS packets, however the common approach is to encapsulate 7 TS packets into one RTP packet. Together with some protocol headers, the packet size stays just below the MTU (Maximum Transmission Unit) of 1500 bytes (Ethernet), which avoids fragmentation. The problem of such RTP packets is that they are dropped by IFD as a whole, and they may contain audio, video and data TS packets, or packets belonging to multiple frames. Therefore we need to make sure that video and non-video packets are split up in different RTP packets, and the same goes for video packets belonging to different frames. Re-ordering of TS packets is unwanted due to timing dependencies and efficiency (e.g. buffer memory usage) considerations, therefore we can achieve the splitting by simply finalizing packets earlier. During parsing, TS packets are gathered into RTP packets. A RTP packet is finalized if one of the following conditions is met:

1. The RTP packet contains 7 TS packets.
2. The next TS packet has a different TS PID (Packet Identifier) than its predecessor, and one of them has the video PID.
3. The next TS packet and its predecessor both have the video PID, but the next packet starts a new video frame.

Figure 2(a) shows the result of the common encapsulation scheme for an example TS packet sequence. The TS packets are denoted with their type (**Audio, Video, Data**). For video packets also their frame numbers are mentioned. All packet boundaries are exactly 7 TS packets apart. Figure 2(b) shows the packet boundaries obtained with our encapsulation scheme. Boundary number 2 is the result of the first rule (maximum of 7 packets reached). Boundaries 1, 3, 4, 6 and 7 are the result of rule 2 (switching between video and non-video). Boundary 5 is the result of crossing a frame boundary (rule 3).

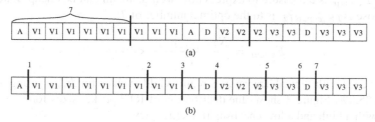

Fig. 2. Example of RTP packet encapsulation: (a) common scheme, (b) IFD-friendly scheme

Compared to the common encapsulation scheme, parsing of the TS packets down to the video elementary stream level is required for finding the picture start codes in video packets to identify the start of each frame and the frame type. In many cases the start of a new frame is aligned with the start of a TS packet. This is the ideal situation and a clean separator can be placed between two TS packets to separate two frames. However, there are cases where the frame start and TS packet start are not aligned:

1. In some cases, the picture start code is located halfway a TS packet, which means that this packet belongs to two different frames. In this case there are a number of options for setting the RTP packet boundary: 1) split up this packet into two TS packets and add stuffing, 2) keep this packet with the previous frame, 3) keep this packet with the new frame. For efficiency reasons (no increased bit-rate and no copying/modification of TS headers) we did not go for option 1. Options 2 and 3 differ in which part of a frame will be dropped. With option 2 the start of the new frame will be lost if the previous frame is dropped, while with option 3 the last part of the previous frame will be lost if the new frame is dropped. In our experience, most decoders handle the latter situation better, therefore we choose option 3.
2. Sometimes the picture start code is split over two TS video packets. Detecting this situation entails a higher complexity, because two video packets have to be scanned at a time, with the additional possibility that they are interleaved with audio and data

packets. Failing to detect this situation may result in two frames being tagged as if they are one. If a B-frame is followed by a P-frame, then also the latter is tagged as a B-frame, and may be dropped as well.

The occurrence of the above situations depends on how the video stream is packetized into Packetized Elementary Stream (PES) packets, before it is multiplexed in the Transport Stream together with audio data. From our experience with broadcast streams two methods are most often used: 1) each PES packet contains a single frame, 2) each PES packet contains a GOP. The above miss-alignment situations do not occur when a PES packet contains a frame, because the start of a PES packet is always at the start of a TS packet. They might occur often, though, when each PES packet contains a GOP.

IFD Overhead. As can be seen from Figure 2, our IFD-friendly encapsulation scheme results in more and smaller RTP packets. Smaller packets have more overhead, which means that the wireless network is used less efficiently. Considering an average size of a RTP packet consisting of $N_{TS\_per\_RTP}$ TS packets, we define the *encapsulation efficiency* E_{encap} as a measure to express how well a stream can be encapsulated, that is, how close $N_{TS\_per\_RTP}$ is to the optimal number of 7:

$$E_{encap} = \frac{N_{TS\_per\_RTP}}{7} \times 100\% \tag{1}$$

Our experience with typical Transport Streams yields values for E_{encap} ranging from 37.4% to 85.8%. Figure 3 shows the distribution of RTP packet sizes for example sequences with a high and a low encapsulation efficiency.

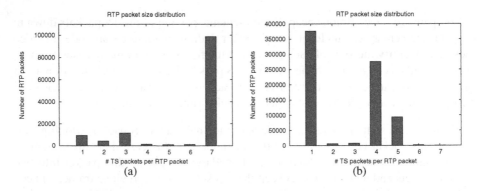

Fig. 3. Example RTP packet size distribution for high (a) and low (b) encapsulation efficiency

One source of overhead of having smaller packets is the protocol headers (e.g. RTP, IP), as the packet payload is relatively smaller. An analysis of the header overhead resulted in numbers ranging from 0.92% to 8.42% for our example streams. Another source of overhead comes from the fact that the maximum theoretical throughput at the 802.11 MAC layer degrades with decreasing packet size (a theoretical analysis can be found in [13]). Figure 4 depicts the relation between $N_{TS\_per\_RTP}$ and the maximum throughput. We define the throughput penalty as the relative drop in throughput with an average RTP packet size $N_{TS\_per\_RTP}$ compared to the throughput with the maximum packet size of 7 ($E_{encap} = 100\%$). For the example sequences we used, a throughput penalty was found ranging from 6.1% to 47%. It is therefore imperative to obtain a high encapsulation efficiency. This depends on how the packets have been multiplexed. The best is to have as few transitions between video and non-video packets as possible. Since video packets make up most of the stream, we can then achieve the highest possible number of RTP packets of the maximum size (such as in Figure 3(a)).

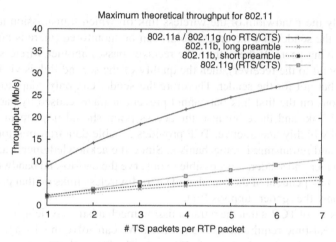

Fig. 4. Maximum theoretical throughput for 802.11

Implementation. We implemented IFD as part of a RTP/RTSP server application running on a Linux PC, the streaming part of which is shown in Figure 5. The *TS file reader* reads TS packets from a MPEG-2 Transport Stream file and sends it to the *RTP packetizer*, which encapsulates the TS packets into RTP packets according to the method described earlier. The target transmission times (TTT) of the packets are determined by the file reader from the PCR (Program Clock Reference) values (following the DLNA guidelines [14]). The RTP packetizer then translates the TTT to RTP timestamps and inserts the proper RTP headers. The resulting RTP packets are given to the *RTP sender*. This component is responsible for buffering the RTP packets for transmission. For the right pacing of the transmission, the *RTP scheduler* examines the RTP timestamps of the RTP packets and tells the RTP sender to send the packets at the right time. IFD packet tagging is done by the RTP packetizer, and the dropping is implemented in the RTP sender. Audio and data packets are tagged with a priority number which is not recognized as video frames by IFD; they thus will never be dropped. In this way the audio is almost never interrupted at the receiver output.

3.3 TCP Implementation

The main drawback of using RTP streaming is that IFD must be supported by each wireless node (e.g. an access point) in the network, not only at the sender. This is because

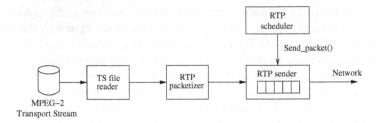

Fig. 5. RTP sender architecture

with RTP, only the bandwidth of the wireless link on which transmission takes place (e.g. between the sender and the access point) can be monitored (there is no feedback from the receiving peer). If the path to the receiver passes another wireless link (e.g. from access point to the receiver), then the quality of the second link is visible for the access point, but not for the sender. Therefore the sender can only adapt to the band-width conditions on the first link but cannot prevent artifacts caused by packet losses on the second link, and therefore also the access point should implement IFD. This is commercially highly unattractive. TCP provides reliable data transmission using its acknowledge and retransmission mechanism. Since the acknowledgments are sent be-tween the receiver and sender, it is possible to observe the end-to-end bandwidth on the network path, independent of the (number of) wired/wireless links on that path. Thus it suffices that only the sender supports IFD.

The main issue of TCP is that the retransmission mechanism may delay the streaming such that the real-time requirements are not met. We can solve this by applying IFD. Our TCP sender architecture is shown in Figure 6. The file reader component is similar to the RTP implementation. The TS collector component packs TS packets together in a similar way as the RTP packetizer described in Section 3.2, with a slight difference that the TS collector never applies encapsulation rule 1. This is because packets offered to TCP may be bigger than 7 TS packets; TCP will automatically split up such big packets into smaller chunks. The IFD algorithm is implemented by the *IFD dropper* component. It writes packets into the IFD buffer according to the pacing of the stream. The *TCP sender* task tries to send the packets in this buffer as fast as possible. When the network conditions deteriorate, TCP stalls and the TCP sender cannot empty the IFD buffer fast enough. This will trigger the IFD dropper to apply the dropping algorithm.

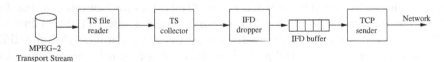

Fig. 6. TCP sender architecture

The overhead introduced by our packetization scheme for TCP is smaller than for RTP. This is because TCP will automatically split up big packets into smaller chunks, and merge smaller packets into bigger chunks for transmission. The average packet size at the IP level is therefore bigger than for RTP. This comes at a cost of a bigger receiver buffer to deal with the higher jitter.

4   Experimental Results

This section presents some experimental results in two scenarios. The first one involves a stationary receiver attached to a TV. In this scenario (Section 4.1), the wireless network is disturbed by turning on a microwave. The second scenario (Section 4.2) involves a mobile receiver, which is moved away from the access point, causing the bandwidth to drop. We used a 802.11g wireless access point and PCMCIA wireless adapters. IFD is implemented in the sender. The experiments were conducted in an office environment with a WLAN infrastructure, which causes some additional interference with our own network.

4.1   Stationary Receiver with Microwave Disturbance

The set-up for this scenario is shown in Figure 7. The sender PC has a wireless connection to the access point, which is connected via wired Ethernet to a set-top-box. The nearby (at appr. 2 m) microwave introduces the disturbance. For this experiment we used a 10 Mb/s sequence. The GOP size is 15, the GOP structure is IBBP. During the

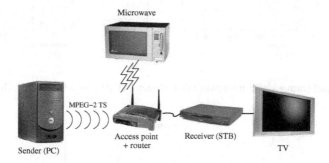

Fig. 7. Experimental set-up with stationary receiver

experiment, the microwave was turned on 45 seconds after the stream started, for a period of 1 minute, and the output video and audio quality was observed on the TV. With the microwave off, the audio and video were streamed without problems. During the period when the microwave was on, the experiment without IFD exhibited artifacts in the video and frequent interruptions in the audio. With IFD turned on, the audio was never interrupted, and the video frame rate was observably reduced, however no artifacts were seen. Similar observations were made both for the RTP and the TCP implementation. Since the wireless network conditions may vary over time, the experiments were repeated a number of times.

Measurements were performed to determine the dropped frames and their types. We first ran an experiment with the TCP sender (Figure 6) without IFD, where the IFD dropper was configured to drop all incoming packets when the IFD buffer is full (tail-drop). The dropped frames at the sender side are shown in Figure 8(a). As can be seen, almost no frames were dropped when the microwave was off. When the microwave was on, frames from all types were dropped (including I- and P-frames, consistent with the observed artifacts). Note that actually only some packets belonging to these frames were dropped, and not the complete frames. We consider a frame with one or more packets missing as dropped because typical decoders will throw away incomplete frames anyway (including I- and P-frames, leading to artifacts). With IFD turned on, it can be seen that no I-frames were dropped (Figure 8(b)). Most of the dropped frames were B-frames. It can be seen that also some P-frames had to be dropped, causing the dropping of the rest of the disturbed GOPs. Considering our GOP size, dropping a disturbed GOP may cause the image to be frozen up to roughly half a second. This effect is reduced if the GOP size is smaller. The results for the RTP implementation are shown in Figure 9. As can be seen, in this experiment only B-frames were dropped.

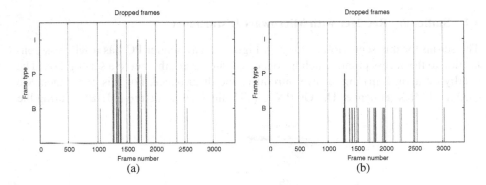

Fig. 8. Dropped frames under microwave disturbance (TCP): (a): without IFD, (b) with IFD

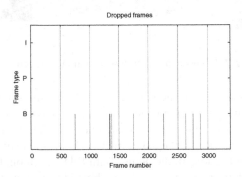

Fig. 9. Dropped frames with IFD under microwave disturbance (RTP)

4.2 Mobile Receiver

In this experiment we used a 8 Mb/s sequence. The sequence was streamed along two wireless links from the sender PC to the laptop (see Figure 10). After the stream started, we walked away with the laptop from the access point until halfway the sequence, then turned around and walked back to the starting position. This experiment was only done with the TCP implementation, because for RTP we lacked IFD on the access point for adapting the stream on the second link.

Fig. 10. Experimental set-up with mobile receiver

With IFD off, it was observed that while walking away, the audio and video were first streamed with no problems, followed by a short period with artifacts, then followed again by a period without problems. This is caused by the automatic reduction of the transmission rate by the access point when the signal strength decreased. Hence the artifacts were caused during the transitions. This effect can also be seen in Figure 11(a) from the short bursts of dropped frames. When we reached a certain distance where the required bandwidth could not be sustained anymore, the receiver output exhibited continuous artifacts. With IFD on, we observed that during transitions the image showed a short freeze (caused by dropping a P-frame and the subsequent frames in the disturbed GOP). There were no blocking artifacts and no audio interruptions. As can be seen from Figure 11(b), the dropped frames were limited to B-frames and P-frames.

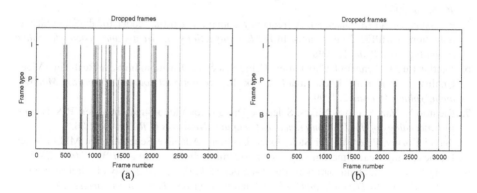

Fig. 11. Dropped frames with roaming receiver: (a) without IFD, (b) with IFD

5 Conclusions

In this paper, we have shown a method for adaptive streaming of audio/video content over wireless networks while using standard content distribution formats and network protocols. With I-Frame Delay as the underlying bit-rate adaptation mechanism, we are able to stream artifact-free video even under degrading network conditions. Our solution is implemented at the sender only, no modifications are needed at the receiver side. It is possible to apply IFD in combination with RTP as well as TCP. We proposed a packet encapsulation scheme which makes it possible to separate video TS packets from non-video packets and packets belonging to different frames, such that the resulting packets can be fed to the IFD scheduler. The effectiveness of IFD in case of network bandwidth fluctuations was shown by means of measurements on a real-life demonstrator set-up.

Our RTP packet encapsulation scheme results in on average smaller network packets, which entails some overhead in network efficiency. This overhead is dependent on the encapsulation efficiency of the streams. For TCP this overhead is considerably smaller.

IFD is effective against bandwidth fluctuations, which may be severe but only last a short period of time. In case of long-term bandwidth drops (e.g. due to multiple contending streams), the perceived quality of the IFD solution seriously degrades because frames are constantly dropped. Such bandwidth problems are better handled by solutions such as transrating.

References

1. D. Wu, Y. Hou and Y.-Q. Zhang. Scalable Video Coding and Transport over Broadband Wireless Networks. In *Proceedings of the IEEE, vol. 89, no. 1*, 2001.
2. M. Domanski, A. Luczak and S. Mackowiak. Spatio-temporal Scalability for MPEG Video Coding. In *IEEE Transactions on Circuits and Systems for Video Technology, vol. 10*, 2000.
3. H. Radha, M. van der Schaar and Y. Chen. The MPEG-4 Fine-Grained Scalable Video Coding Method for Multimedia Streaming over IP. In *IEEE Transactions on Multimedia, vol. 3, no. 1*, 2001.
4. Kwang-deok Seo, Sang-hee Lee, Jae-kyoon Kim and Jong-seog Koh. Rate-control algorithm for fast bit-rate conversion transcoding. In *IEEE Transactions on Consumer Electronics, vol. 46, no. 4*, 2000.
5. P. Assuncao and M. Ghanbari. A frequency-domain video transcoder for dynamic bit-rate reduction of MPEG-2 bit streams. In *IEEE Transactions on circuits and systems for video technology, vol. 8, no. 8*, 1998.
6. Yantian Lu and Kenneth J. Christensen. Using Selective Discard to Improve Real-Time Video Quality on an Ethernet Local Area Network. In *International Journal of Network Management, Volume 9, Issue 2*, 1999.
7. Ricardo N. Vaz, Mário S. Nunes. Selective Discard for Video Streaming over IP Networks. In *Proceedings of the 7th Conference on Computer Networks (CRC2004)*, 2004.
8. Sergei Kozlov, Peter van der Stok, Johan Lukkien. Adaptive Scheduling of MPEG Video Frames during Real-Time Wireless Video Streaming. In *Proceedings of WoWMoM*, 2005.
9. Reinder Haakma, Dmitri Jarnikov, Peter van der Stok. Perceived quality of wirelessly transported videos. In *Dynamic and Robust Streaming in and between Connected Consumer-Electronic Devices*, Philips Research Book Series, vol. 3. Springer, 2005.
10. SangHoon Park, Seungjoo Lee, JongWon Kim. Network-adaptive high definition MPEG-2 streaming over IEEE 802.11a WLAN using frame-based prioritized packetization. In *Proceedings of the 3rd ACM international workshop on Wireless mobile applications and services on WLAN hotspots (WMASH)*, 2005.
11. ISO/IEC 13818-1:2000(E). Information technology − Generic coding of moving pictures and associated audio information: System.
12. D. Hoffman, G. Fernando, V. Goyal and M. Civanlar. RFC2250: RTP Payload Format for MPEG1/MPEG2 Video.
13. Jangeun Jun, Pushkin Peddabachagari and Mihail Sichitiu. Theoretical Maximum Throughput of IEEE 802.11 and its Applications. In *Proceedings of the 2nd IEEE International Symposium on Network Computing and Applications (NCA'03)*, 2003.
14. Digital Living Network Alliance. Home Networked Device Interoperability Guidelines, Version 1.5, 25-10-2005.

Management of IEEE 802.11e Wireless LAN for Realtime QoS-Guaranteed Teleconference Service with Differentiated H.264 Video Transmission*

Soo-Yong Koo, Byung-Kil Kim, and Young-Tak Kim[**]

Dept. of Information and Communication Engineering,
Graduate School, Yeungnam University
214-1, Dae-Dong, Kyungsan-Si, Kyungbook, 712-749, Korea
sykoo@tsc.ac.kr, woo_sue@hotmail.com, ytkim@yu.ac.kr

Abstract. Various realtime multimedia applications will be provided in next generation Internet where IEEE 802.11e wireless LAN will be widely used as broadband access networks. The characteristics of the wireless channels in IEEE 802.11 (i.e., fluctuating bandwidth and large error rate), however, impose challenging problems in the efficient QoS-guaranteed realtime multimedia communications with strict QoS requirements (i.e., bandwidth, delay, jitter, and packet loss/error rate). In this paper we propose management schemes of IEEE 802.11e wireless LAN (WLAN) for realtime QoS-guaranteed teleconference services with differentiated H.264 video transmission. In the proposed scheme, the IEEE 802.11e Wireless LAN is managed to transmit I, P and B slices from H.264 encoder using different channels of both HCF controlled channel access (HCCA) and enhanced distributed channel access (EDCA). We compare several different mapping scenarios, and analyze the QoS provisioning performance for realtime multimedia teleconference service.

Keywords: IEEE 802.11e, H.264, QoS, Diffserv, HCCA, EDCA.

1 Introduction

End-to-end QoS-guaranteed differentiated service provisioning is essential in next generation Internet and Broadband convergence Network (BcN) where various wired and wireless networks are interconnected. Especially, realtime multimedia services, such as VoIP and multimedia teleconference, require strict bandwidth and QoS requirements (i.e., committed data rate with given burst size, delay, jitter, and packet loss/error rate). The IEEE 802.11 wireless LAN [1-4] and IEEE 802.16 wireless MAN are expected to be used widely in next generation Internet as broadband wireless access network. The characteristics of the wireless channel in IEEE 802.11 (i.e., fluctuating bandwidth and large error rate), however, impose challenging

---

[*] This research was supported by the MIC, under the ITRC support program supervised by the IITA.

[**] Corresponding author.

A. Helmy et al. (Eds.): MMNS 2006, LNCS 4267, pp. 25–36, 2006.

problems in the efficient QoS-guaranteed realtime multimedia communications with strict QoS requirements.

H.264/AVC is the newest video coding standard that was approved by ITU-T as Recommendation H.264 and by ISO/IEC as International Standard 14496-10(MPEG-4 part 10) Advanced Video Coding (AVC)[5-8]. H.264/AVC provides good video quality at broad range of bit rates and picture sizes, ranging from very low bit rate, low frame rate, postage stamp resolution video for mobile and dial-up devices, through to entertainment-quality standard definition(SD) and high definition(HD) TV. H.264/AVC also can be efficiently used in a resource limited environments such as WLAN. H.264 video encoder generates video stream in separate units of I (intra), P (predicted), and B (bi-predictive) slices according to the selected profile (baseline, main, extended) for different applications.

IEEE 802.11e [1] proposed new enhanced QoS provisioning mechanism that promises to ensure good QoS to applications depending upon its traffic category & type. These two channel access functions are managed by a centralized controller called Hybrid Coordinator (HC) which is a module in the QoS Access Point (QAP).

Most research works for video transmission over IEEE 802.11e have been studied with EDCA mechanism[9]. In [9], where parameter set information (PSI), instantaneous decoding refresh (IDR) picture slice, partition A, partition B, partition C are transmitted through EDCA access category AC_VO, AC_VI and AC_BE. The differentiated video transmission with delay constraints for teleconference on IEEE 802.11e WLAN with HCCA and EDCA, however, has not been fully studied yet. In this paper, we analyze the management of the IEEE 802.11e wireless LAN channels of HCCA and EDCA for QoS-guaranteed realtime multimedia teleconference service with differentiated H.264/AVC video transmission.

The rest of this paper is organized as follows. In section 2, we briefly explain the related work, such as H.264/AVC, IEEE 802.11e with HCCA and EDCA and transmission of H.264 video traffic over IEEE 802.11e. In section 3, we study various mapping scenarios of H.264/AVC video transmission on IEEE 802.11e HCCA & EDCA channels. In section 4, we analyze the QoS provisioning performance of each mapping scenario, considering the end-to-end delay (less than 400 ms) for multimedia teleconference. Finally we conclude in section 5.

2 Background and Related Work

2.1 IEEE 802.11e Wireless LAN with HCCA and EDCA

IEEE 802.11e [1] proposed new enhanced QoS provisioning mechanism that promises to ensure good QoS to applications depending upon its traffic category & type. IEEE 802.11e MAC includes an additional coordination function called Hybrid Coordination Function (HCF). The HCF uses both a contention-based channel access method, called the enhanced distributed channel access (EDCA) mechanism for contention-based transfer, and a controlled channel access, referred to as the HCF controlled channel access (HCCA) mechanism, for contention-free transfer[1].

The QoS provisioning on IEEE 802.11e is based on enhanced distributed channel access by EDCA and centralized channel access by HCCA. These two channel access functions are managed by a centralized controller called Hybrid Coordinator (HC) which is a module in the QoS Access Point (QAP).

EDCA is the contention-based medium access method, and is realized with the introduction of traffic categories (TCs). The EDCA provides differentiated distributed access to the wireless medium for 8 priorities of stations. EDCA defines the access category (AC) mechanism that provides support for the priorities at the stations. Each station may have up to 4 ACs (AC_VO, AC_VI, AC_BE and AC_BK) to support 8 user priorities (UPs). One or more UPs are assigned to one AC. Even though EDCA provides differentiated access categories, it does not guarantee the QoS parameters of hard realtime applications, i.e., jitter and delay.

In order to provide realtime services with guaranteed QoS-parameters, HCCA that has been designed for parameterized QoS support with contention-free polling-based channel access mechanism must be used. The QAP scheduler computes the duration of polled-TXOP (transmission opportunity) for each QSTA based upon the traffic specification (TSPEC) parameters of an application flow. The scheduler in each QSTA then allocates the TXOP for different traffic stream (TS) queues according to the priority order. In IEEE 802.11e, TSPEC is used to describe the traffic characteristics and the QoS requirements of a data flow to and from QSTA.

2.2 H.264/AVC Standard

H.264/AVC has been developed for higher compression of moving pictures for various applications such as videoconferencing, digital storage media, television broadcasting, Internet streaming, and communication [5]. It is also designed to enable the use of the coded video representation in a flexible manner for a wide variety of network environments. H.264/AVC introduces a set of error resiliency techniques such as slice structure, data partitioning (DP), flexible macroblock ordering (FMO), arbitrary slice ordering (ASO), and redundant pictures.

H.264/AVC is divided into two distinct layers. First, the video coding layer (VCL) is responsible for efficient representation of the video data based on motion compensation, transform coding, etc. Second, the network abstraction layer (NAL) is responsible for delivery over various types of network. H.264/AVC codec maps VCL data (a sequence of bit representing the coded video picture) into packets known as NAL units (NALUs) prior to transmission or storage. An NALU corresponds to a slice (or a parameter set). Each slice is to be packetized into its own RTP packet conforming to the RFC 3984 [12] packetization scheme. An NALU consists of a one-byte header and raw byte sequence payload (RBSP) that represents the MBs of a slice.

H.264/AVC defines three profiles, each supporting a particular set of coding functions for potential different applications. *Baseline profile* supports intra and inter-coding (using I-slices and P-slices), and its potential applications include video-telephony, video-conference, and wireless communications. *Main profile* supports interlaced video, inter-coding using B-slices, inter-coding using weighted prediction and entropy coding, and its potential application includes television broadcasting and video storage. *Extended profile* does not support interlaced video, but adds modes to enable efficient switching between coded bit streams and improved error resilience (i.e. data partitioning), and is particularly useful for streaming media applications.

2.3 Transmission of H.264 over IEEE 802.11e

Adlen Ksentini et. al proposed a cross-layer architecture of H.264 video transmission over IEEE 802.11e WLAN that leverage the inherent H.264 error resilience tools and the existing QoS-based IEEE 802.11e MAC protocol possibilities [9]. In this cross-layer design, authors favor more interaction between the H.264's VCL that divides the original streams through data partitioning (DP) and the MAC that treats video streams with different EDCA access categories. NAL encapsulates slices generated by VCL to NALU. [9] focuses particularly on the nal_ref_idc (NRI) field in the NAL header. The NRI contains two bits that indicate the priority of the NALU payload, where 11 is the highest transport priority, followed by 10, then by 01, and finally, 00 is the lowest.

In the proposed architecture, the slices from the VCL have been differentiated according to the slice type (i.e., parameter set information, IDR picture, partition A, B and C). EDCA access category 3 (AC_VO) is used to transmit parameter set information, while EDCA AC_VI is used to transmit IDR picture and partition A. EDCA AC_BE has been used to transmit partition B and partition C. In this way, partition B and C are differentiated from background traffic (AC_BK).

In the proposed architecture, H.264 extended profile has been used with data partitioning option. The results obtained indicated that the proposed architecture achieves better performances in terms of delays and loss rate that the IEEE 802.11 DCF and 802.11e EDCA with single access category. The packet delays of IDR and partition A of the proposed architecture are shown to be less than 500 ms; however, the packet delays of partition B and C are not explained, and the overall PSNR is not analyzed. Also, the proposed mapping algorithm uses only EDCA access categories to transmit H.264 packets. It may cause higher packet loss in high priority ACs and degradation of perceived video quality when network load becomes significant, or many nodes that transmit H.264 traffic compete with each other.

3   H.264/AVC Video Transmission on IEEE 802.11e HCCA and EDCA Channels

3.1   Analysis of H.264/AVC Video

In order to design an efficient mapping architecture of H.264/AVC video slices/ partitions and IEEE 802.11e HCCA/EDCA channels, we firstly analyzed the amount of traffic generated of each slice type, and the relationship between packet loss rate of each slice type and the peak signal to noise ratio (PSNR) in extended profile and baseline profile, respectively. For the analysis, Foreman and Paris CIF(352x288) video sequence have been used, these sample video clips were encoded using JM 10.2 reference software[13] with following encoder parameters setting: IDR picture period is 30 frames, frame rate is set constant at 30fps, output file mode is RTP, and the slice size is 1500bytes.

Table 1. Composition of slice types in the results of H.264 encoding

(a) Case 1 – Extended profile

		PSI	I	P	B	Total
Foreman	percentage	0.002%	10.25%	75.36%	14.388%	100%(394kbps)
(9.8sec)	# of pkts	2	36	320	149	507
Paris	percentage	0.002%	20.008%	66%	13%	100%(385kbps)
(35sec)	# of pkts	2	238	1019	531	1790

(b)Case 2 – Baseline profile

		PSI	I	P	Total
Foreman	percentage	0.002%	16.92%	83.08%	100%(474kbps)
(9.8sec)	# of pkts	2	73	470	545
Paris	percentage	0.002%	29.008%	70%	100%(503kbps)
(35sec)	# of pkts	2	474	1543	2019

Table 1 (a) and (b) show the composition of the slice types in H.264 encoding with extended profile and baseline profile, respectively. In extended profile, the total transmission rates are 394 kbps for Foreman sequence and 385kbps for Paris sequence, where generated traffic of I slice is 10.25 %, P slices 75.36 %, and B slice 14.387% for Foreman sequence in extended profile. In baseline profile, only I and P slices are generated, and the ratio of PSI, I, and P slices for Foreman sequence are 0.002%, 16.92%, and 83.08% respectively. The compositions of the slice types of Paris sequence have been analyzed to be in the same trend.

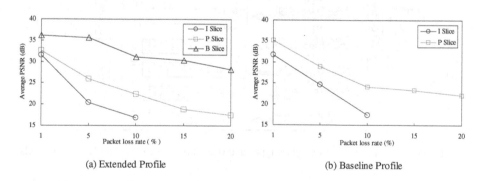

(a) Extended Profile   (b) Baseline Profile

Fig. 1. PSNR according to packet loss rate of each H.264 slice type

Fig. 1 depicts the relationship between packet loss rate of each H.264 slice type and the peak signal to noise ratio (PSNR) for Foreman sequence. Packet loss in I slice has the highest impact on the PSNR, P slice has the next impact, while B slice has the least impact on PSNR. This analysis results mean that I slice should be given the highest priority for least packet loss, while B slice can be given the lowest priority among the PSI, I, P and B slices from H.264 encoder. In the H.264 decoding simulation in ns-2, we found that when the PSNR is less than 30 dB, the result video replay was at the status of annoying. So, the packet loss rate of I slices and P slices must be guaranteed to be less than 5 %.

3.2 Differentiated QoS-Provisioning with IEEE 802.11e HCCA and EDCA Channels

Fig. 2. depicts a mapping scheme between DiffServ class-types and IEEE 802.11e HCCA and EDCA channels. Since the class-types of NCT, EF and AF4 require tight end-to-end packet delay and jitter bound, they are mapped onto HCCA; AF3/2/1 and BEF are mapped onto EDCA that can support flexible bandwidth usage with less stringent time constraints. By allocating HCCA channels, we can guarantee the bandwidth, while some EDCA channels may not guarantee bandwidth and delay if admission control mandatory (ACM) is not configured in lower priority channels.

In the differentiated transmission in IEEE 802.11e HCCA and EDCA channels, the delay, jitter, and available bandwidth guarantee are very important in the QoS-guaranteed multimedia service provisioning. By default, the overall available bandwidth of HCCA channels is limited to 40% of the total available physical layer bandwidth. The 4 access categories (AC_VO, AC_VI, AC_BE and AC_BK) in EDCA provides differentiated channel access priority among different access categories, but if multiple channel accesses are requested for a same access category, contentions are generated, and guaranteed bandwidth provisioning is not possible.

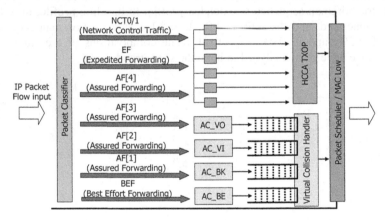

Fig. 2. Mapping between DiffServ class-types and IEEE 802.11e HCCA and EDCA channels

In the mapping scenarios between H.264 slice type and IEEE 802.11e HCCA and EDCA channels, we can consider several different alternatives, as shown in Table 2. In H.264 baseline profile encoding, only I and P slices are generated without B slices.

Table 2. Alternatives in mapping H.264 slice type and IEEE 802.11e HCCA and EDCA

	PSI	I	P	B
Case 1 (AC_VI only)	AC_VI	AC_VI	AC_VI	AC_VI
Case 2 (AC_VO+AC_VI)	AC_VO	AC_VO	AC_VI	AC_VI
Case 3 (HCCA+ AC_VI)	HCCA	HCCA	AC_VI	AC_VI
Case 4 (AC_VO+AC_VI+AC_BE)	AC_VO	AC_VO	AC_VI	AC_BE
Case 5 (HCCA+ AC_VI+AC_BE)	HCCA	HCCA	AC_VI	AC_BE

So, only case 1 ~ 3 are applicable for H.264 baseline profile encoding for video phone and teleconference.

4 Performance Analysis of Differentiated H.264 Video Transmission on IEEE 802.11e EDCA and HCCA Channels

4.1 Simulation Configuration

In order to evaluate the differentiated QoS provisioning performance of IEEE 802.11e with HCCA and EDCA, we performed a series of ns-2 simulations. In our simulation, the QSTA are communicating through IEEE 802.11b physical wireless links with 6Mbps physical transmission rate in infrastructure mode.

Currently, in the ns-2 network simulation configuration, each QSTA generates same H.264/AVC video traffic without internal virtual collision detection. To verify the performance of QoS provisioning, we measured the throughput, delay, jitter and packet loss rate. To generate network congestion situation, the number of active QSTAs has been gradually increased, and the performance parameters have been measured. At a certain level of increased traffic amount, contention occurs and severe packet drop and increased delay deteriorate the video quality.

4.2 Analysis of the H.264 Extended Profile Video Transmissions

Fig. 3, Fig. 4 and Fig. 5 depicts the throughput, packet delay and packet loss in the mapping scenarios for H.264 extended profile video transmission where I, P and B slices are transmitted.

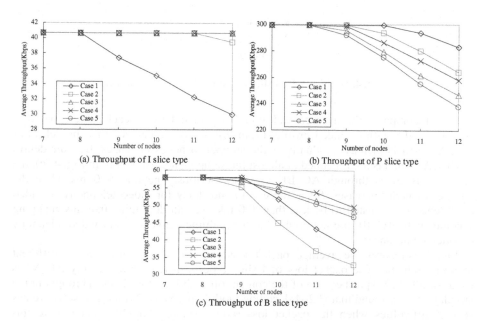

(a) Throughput of I slice type (b) Throughput of P slice type

(c) Throughput of B slice type

Fig. 3. Throughput of I, P, B slice types from H.264 extended profile

Fig. 3 compares the average throughputs of each slice type at different mapping scheme. In Fig. 3 (a), when I slices are transmitted via HCCA or AC_VO (Case 2 ~ 5), the throughput is guaranteed even when the number of node increases; when I slices are transmitted via AC_VI only (Case 1) competing with other P and B slices, the throughput is reduced as the number of node increases. These trends are also shown in Fig. 3 (b) where P slices are transmitted through AC_VI or AC_VO competing with other nodes or other slice types. In Fig. 3 (c), we can see that when B slices are transmitted through AC_BE competing with other slice types and nodes, its throughput is reduced seriously.

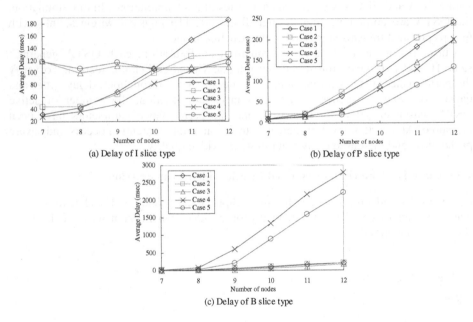

(a) Delay of I slice type

(b) Delay of P slice type

(c) Delay of B slice type

Fig. 4. Delay of I, P, B slices from H.264 extended profile

Fig. 4 compares the average end-to-end packet delays of each slice type from H.264 encoder with extended profile. From Fig. 4 (a), we can see that the delays of EDCA channels increase gradually as the number of nodes increases. But, the delays of HCCA channels are remained at almost the same value. Especially in Fig. 4 (c), the delay of B slices through AC_BE increases seriously beyond 500 ms when the number of node increases. Since the end-to-end delay for video telephony or video conference is requested to be less than 400 ms, we can determine that any mapping schemes with AC_BE are not applicable for realtime conversational video telephony or video conference.

Fig. 5 compares the average packet loss ratios of each slice type at different mapping scheme. The packet loss of I slice type is increased when only AC_VI is used for all slice types because of the contention with P and B slices. Throughout the simulations, we found that H.264 decoding module in JM 10.2 reference software has severe difficulties when the packet loss ratio is more than 20 %. To solve this decoding problems, appropriate error concealment functions are needed.

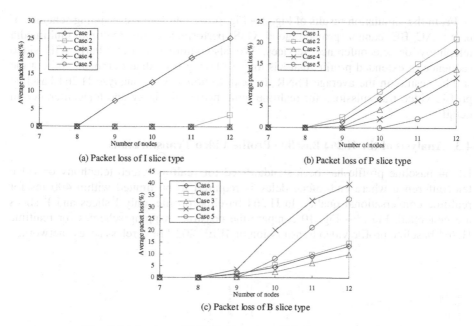

(a) Packet loss of I slice type

(b) Packet loss of P slice type

(c) Packet loss of B slice type

Fig. 5. Packet loss of I, P, B slices from H.264 extended profile

Fig. 6 depicts the comparison of average PSNRs of each mapping scheme. The mapping schemes of using Case 1(AC_VI only) and Case 4(AC_VO+ AC_VI+AC_BE) are showing severely deteriorated PSNR less than 30 dB when the number of nodes is increased beyond 11. But, the mapping scheme that uses HCCA for I slices, AC_VI for P slices, and AC_BE for B slices (Case 5) maintains the acceptable value of PSNR according to the number of node increases, because high priority packets (I and P slice packets) experience less packet loss than other schemes. In Case 5, packet loss rate of B slice type is more severe but less influence on PSNR as previously mentioned in chapter 3.1 (Fig. 1).

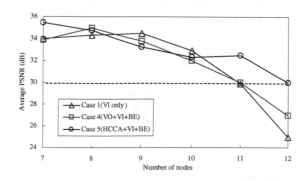

Fig. 6. Comparisons of PSNRs of mapping schemes for H.264 extended profile

From the simulation results of Fig. 3 ~ Fig. 6, we can find the mapping schemes of using AC_BE cannot provide good QoS performance for realtime multimedia telephony or teleconference services. We also simulated using Paris CIF video sequence in extended profile, that has longer playing time than Foreman, and got the similar results in the average PSNR. In next section 4.3, we analyze H.264 baseline profile video transmissions for realtime multimedia telephony and teleconference in detail.

4.3 Analysis of the H.264 Baseline Profile Video Transmissions

H.264 baseline profile has been standardized for realtime video telephony or video teleconference where end-to-end delay is required to be limited within 400 ms for realtime conversational quality. In H.264 baseline profile, only I slices and P slices are generated. Fig. 7 ~ Fig. 10 compare the possible mapping schemes for realtime H.264 baseline profile video transmission on IEEE 802.11e wireless access network.

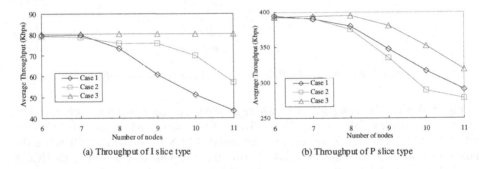

(a) Throughput of I slice type   (b) Throughput of P slice type

Fig. 7. Throughput of I and P slices from H.264 baseline profile

In Fig. 7, Fig. 8 and Fig. 9, we can see that the mapping scheme of HCCA for I slices and EDCA AC_VI for P slices (Case 3) provides the best performance in guaranteed throughput, delay and packet loss ratio. The delays are within the limit of 400 ms in our simulation results, which are reasonable for realtime video teleconference in IEEE 802.11e.

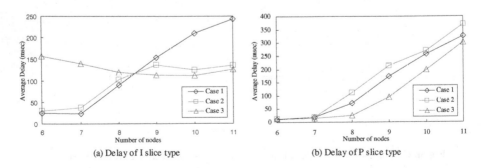

(a) Delay of I slice type   (b) Delay of P slice type

Fig. 8. Delay of I and P slices from H.264 baseline profile

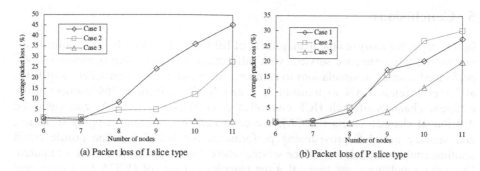

(a) Packet loss of I slice type   (b) Packet loss of P slice type

Fig. 9. Packet loss of I and P slices from H.264 baseline profile

As shown in Fig. 9 (a), Case 1 shows increased packet loss (beyond 25 %) when the number of nodes is beyond 10, and the H.264 decoding at JM 1.0 was not possible. There are no packet losses in Case 3 because HCCA reserves sufficient bandwidth through TSPEC negotiation in connection setup time similar in the previous simulation with extended profile.

Fig. 10 compares the PSNRs of mapping schemes for H.264 baseline profile for realtime video teleconference. The H.264 video transmission using Case 1 (AC_VI only) and Case 2 (AC_VO+AC_VI) show poor performance in PSNR when the number of node increases beyond 8.

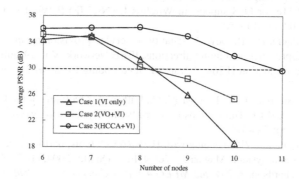

Fig. 10. Comparison of PSNRs of mapping schemes for H.264 baseline profile

From the various simulation analysis on the mapping scheme of H.264 baseline profile on IEEE 802.11e HCCA and EDCA channels for realtime video teleconference application, we can find that the mapping scheme of HCCA for I slices and EDCA AC_VI for P slices of H.264 baseline profile provides the best performance in guaranteed bandwidth provisioning, delay, packet loss and overall PSNR for realtime video teleconference. We also simulated using Paris CIF video sequence in baseline profile, and got the similar results in the average PSNR.

5 Conclusions

In this paper we analyzed the management of IEEE 802.11e WLAN for realtime QoS-guaranteed teleconference services with differentiated H.264 video transmission. We performed a series of simulations to compare the channel allocation schemes of IEEE 802.11e Wireless LAN to transmit I, P and B slices from H.264 encoder using different channels of both HCF controlled channel access (HCCA) and enhanced distributed channel access (EDCA). We compared the different mapping scenarios, and analyze the QoS provisioning performance for H.264 baseline profile based realtime multimedia teleconference service where 400 ms delay time limit is required. From the simulations, we found that the mapping scheme of HCCA for I slices and EDCA AC_VI for P slices of H.264 baseline profile provides the best performance in guaranteed bandwidth provisioning, delay (less than 400 ms), packet loss and overall PSNR for realtime video teleconference.

References

1. IEEE Standard for Information Technology, Local and metropolitan area networks – Part 11: Wireless Medium Access Control (MAC) and Physical Layer (PHY) specifications, Amendment 7: Medium Access Control (MAC) Quality of Service (QoS) Enhancements, IEEE std 802.11e/D12.0, 2004.
2. Qiang Ni, "Performance Analysis and Enhancements for IEEE 802.11e Wireless Networks," IEEE Network, July/August 2005, pp. 21 ~ 27.
3. Stefan Mangold, Sunghyun Choi, Guido Hiertz, Ole Klein, and Bernhard Walke, "Analysis of IEEE 802.11e for QoS support in Wireless LANs," IEEE Wireless Communications, December 2003, pp. 40 ~ 50.
4. Ramos, N., Panigrahi, D. Dey, S., Quality of service provisioning in 802.11e networks: challenges, approaches, and future directions, IEEE Network, Vol. 19, July-Aug 2005.
5. ITU-T Recommendation H.264, Advanced video coding for generic audiovisual services, March 2005.
6. Wiegand, T., Sullivan, G. J., Bjntegaard, G., Luthra, A, Overview of the H.264/AVC video coding standard, IEEE Transactions on Circuits and Systems for Video Technology, Vol. 13, July 2003, ppl 560-576.
7. Ostermann, J., et. al., Video coding with H.264/AVC: tools, performance, and complexity, IEEE Circuits and Systems Magazine, Vol. 4, First Quarter 2004, pp. 7-28.
8. Iain E. G. Richardson, *H.264 and MPEG-4 Video Compression – Video Coding for Next Generation Multimedia*, John Wiley & Sons, 2003.
9. Adlen Ksentini, Mohamed Naimi, Abdelhak Gueroui, "Toward an improvement of H.264 video transmission over IEEE 802.11e through a cross-layer architecture," IEEE Comm. Mag., January 2006, pp. 107-114.
10. Deyun Gao, Jianfei Cai, Bao, P., Zhihai He, MPEG-4 video streaming quality evaluation in IEEE 802.11e WLANs, ICIP 2005, Vol. 1, Sept. 2005, pp. 197-200.
11. Yang Xiao, Voice and Video Transmissions with Global Data Parameter Control for the IEEE 802.11e Enhance Distributed Channel Access, IEEE Transactions on Parallel and Distributed Systems, Vol. 15, No. 11, Nov. 2004, pp. 1041-1053.
12. S. Wenger, M.M. Hannuksela, T. Stockhammer, M. Westerlund, D. Singer, RTP Payload Format for H.264 Video, Internet proposed standard RFC 3984, February 2005.
13. JM 10.2 reference software, http://iphome.hhi.de/suehring/tml/.

Link Available Bandwidth Monitoring for QoS Routing with AODV in Ad Hoc Networks

Stephane Lohier[1], Yacine Ghamri Doudane[2], and Guy Pujolle[1]

[1] LIP6 - University of Paris VI, 8, rue du Capitaine Scott, 75015 Paris – France
[2] IIE-CNAM, 18, allée Jean Rostand, 91025 Evry Cedex – France
stephane.lohier@lip6.fr, ghamri@iie.cnam.fr,
guy.pujolle@lip6.fr

Abstract. Due to bandwidth constraint and dynamic topology of mobile ad hoc networks, supporting Quality of Service is a challenging task. In this paper we present a solution for QoS routing based on an extension of the AODV reactive routing protocol that deals with bandwidth monitoring. The solution uses an IEEE 802.11 MAC layer as the underlying technology and the QoS routing decision is based on simple but accurate measurements, at the MAC layer, of the available bandwidth on each link of the route. In addition, to allow a QoS loss recovery, a notification mechanism is used to inform the source about bandwidth degradation on a link. This reactive solution using standard protocols is adapted to small and dynamic ad hoc networks. A complete simulation set shows that, with the proposed QoS routing protocol, bandwidth on a route is significantly improved without overhead.

Keywords: QoS routing, ad hoc, AODV, IEEE 802.11.

1 Introduction

Throughputs reached today by mobile ad hoc networks based on the IEEE 802.11b and 802.11g standards [1-2] enable the execution of complex applications (video conference, distribution of multimedia flows...). However, these applications consume significant amounts of resources and can suffer from an inefficient and unfair use of the wireless channel. Therefore, new specific QoS solutions need to be developed taking into account the dynamic nature of ad hoc networks. Since these networks should deal with the limited radio range and mobility of their nodes, we believe that the best way to offer QoS is to integrate it in routing protocols. These protocols will have to take into consideration the QoS required by the applications, such as bandwidth constraints, in order to select the adequate routes.

In our current work, a new available bandwidth monitoring technique is proposed. This monitoring is integrated to the routing process. It acts on each link allowing determining if a particular bandwidth demand can be granted or not. In addition, bandwidth monitoring allows detecting, after the QoS route is traced, if the available bandwidth is degraded on a particular link and thus leading to QoS loss on the route using the concerned link. In that case, a QoS loss recovery mechanism is triggered to inform the source which will be able to start a new QoS route search.

A. Helmy et al. (Eds.): MMNS 2006, LNCS 4267, pp. 37–48, 2006.
© IFIP International Federation for Information Processing 2006

The QoS routing process is based on the AODV (Ad hoc On-demand Distance Vector) routing protocol [3] and uses QoS extensions proposed by the originators of AODV [4]. To that first proposal, we add the bandwidth measurement and the QoS loss recovery mechanism. The choice of a standard and reactive routing protocol such as AODV is justified in small, highly dynamic and mobile ad hoc networks. Indeed, it is proved that reactive protocols are well suited in this kind of networks [5-6] which respond to the most common use cases, such as meetings or emergencies, where the number of nodes is limited and variable in a reduced space.

Furthermore, the bandwidth measurements are realized according to 802.11 operations without influencing them. Theses measures are thus passive and compatible with the reactive routing process. Thus, our solution uses recognized and standard protocols without modify them: no modification of the 802.11 standard for the measurements and QoS routing extensions for AODV without additional signalling.

The rest of the paper is organized as follows. In section 2, we introduce the different solutions for bandwidth monitoring in ad hoc networks and the selected one. Section 3 describes the QoS routing process. Section 4 concerns the QoS loss recovery mechanism. Section 5 presents the simulation results. Finally, section 6 concludes the paper and presents some perspectives.

2   Bandwidth Measurement

Before describing the different solutions for the bandwidth measurement, let us clarify some important preliminary points in an ad hoc context and with a reactive routing protocol:

- As the objective is to establish a QoS route between the nodes, the available bandwidth (abw) must be monitored for each link of the route rather than in the neighbourhood of each node.
- For a given source node A, the estimate on the link can be different according to the receiving node ($abw_{AB} \neq abw_{AC}$). Indeed, the receivers do not have the same neighborhood. Moreover the estimate is not symmetrical because the transmitter and receiver nodes do not have the same neighborhood ($abw_{AB} \neq abw_{BA}$).
- To estimate abw_{AB} on the link between A and B, it is necessary to take into account the bandwidth consumed by A and B neighbors.
- It is also necessary to take into account the disturbances caused by the nodes in the interference zone of the receiver B (it is generally considered that the interference zone has a ray double of that of the neighbourhood zone). Indeed, the transmissions of a node in the interference zone of B may cause an important decrease in the Signal to Noise Ratio (SNR), leading to frame losses in the vicinity of B.

Then, the different studies present in the literature can be classified according to the layer where they act: network or link.

At the network layer, the study [7] presents an estimation method, named *hello method,* which is very close to that proposed in the BRuIT protocol [8], but with its integration into the routing process. The principle is that a node can estimate its own available bandwidth when receiving "hello" probes containing information on the bandwidth consumed by its one hop neighbours (two hops for BRuIT protocol). The principal disadvantage of these type of solutions is that they require an important control traffic, particularly for the messages exchanges intended to maintain important and static routing tables, which gives a proactive character on the QoS routing process, more adapted to dense and not very mobile ad hoc networks. In addition, in [7], the "QoS-aware" approach is slightly different from our insofar as the aim is to preserve the end to end average throughput, compared to the solution without QoS, and not to guarantee the throughput required by a source.

At the link layer, most of proposed solutions [9-11] are based on the calculation of the medium availability ratio for each node by measuring the idle and/or the activity periods at the MAC layer. The main disadvantage of these methods is the need for emitting systematically a consequent number of measurement packets for each estimate, which is not appropriate if we want to keep the reactive nature of the routing strategy. In addition, the method is intrusive as it disturbs the neighbouring flows and thus has an influence on the value which it is supposed to measure.

Finally, the constraints related to the selected context (reactive routing, bandwidth on a link, neighborhood zone and interference zone) show us that the estimates are more relevant at the MAC layer, even if the measure is not passive. Indeed, the consideration of neighboring or interfering flows is intrinsically related to measurement and does not require implementing complex signaling overhead at the network layer, which would be in opposition to the reactive character of the routing.

The various solutions at the MAC layer are all based on an initial measurement of the activity or idle periods of the radio channel. These durations are not directly available on an 802.11 interface but can be measured simply by introducing measurement packets provided with timestamps. It is the method employed in [10] and [11]. We include in our proposal this method which takes directly into account the effects of neighbouring and interfering flows. To this method, we add a correction on the packet size, a mechanism allowing to alert the source of an increase in the bandwidth on the link and we avoid the systematic use of a measurement window, conflicting to the reactive character.

Let us detail the measurement principle at the MAC layer on a link between two stations A and B. The various stages of transmission on the 802.11 channel are summarized on Fig. 1. We are interested here in the duration of the channel occupancy and in the duration d_{AB} between the moment when the packet is ready to be emitted by A at the MAC layer and the moment when the packet is received by B (i.e. the moment when the MAC acknowledgement is received by A).

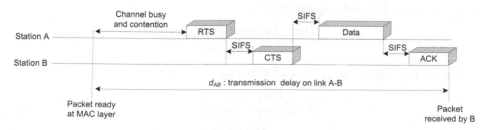

Fig. 1. Transmission stages on the 802.11 channel

A first estimate of *abw* is carried out at the sender side by calculating the relationship between the size of the measurement packet and the d_{AB} duration necessary to its transmission on the channel:

$$abw = \frac{Data\_length}{d_{AB}} \tag{1}$$

We can split up d_{AB} into a variable part and a constant part. The variable part depends on the channel occupancy and on the duration of the contention window. The constant part corresponds to the transmissions of the control and data frames when station A is in emission phase:

$$abw = \frac{Data\_length}{T_{BUSY} + T_{CW} + T_{CST}} \tag{2}$$

The constant term T_{CST} shows dependence with the data packets's size (t_{MPDU}):

$$T_{CST} = t_{RTS} + t_{CTS} + t_{MPDU} + 3t_{SIFS} + t_{ACK} + 4t_{PHY} \tag{3}$$

The duration T_{BUSY} which corresponds to the sequence of the various NAV (*Network Allocation Vector*) timers imposed by the stations in emission, until the station has the right to emit, is directly a function of the traffic in the neighbourhood and interference zones. The duration of the contention window T_{CW} is related to the *backoff* algorithm of 802.11 standard, it is thus a partly random duration but also related on the contentions and the retransmissions and consequently also dependent on the number of neighbouring and interfering flows.

We thus verify that in theory, this first simple estimate of the available bandwidth on the link is related to the sender's and receiver's neighbouring flows (if the contentions increase, T_{BUSY} and T_{CW} increases and *abw* decreases) and also to the flows in the interference zone of the receiver (if the number of retransmissions caused by losses or errors increases, the MAC acknowledgements will be delivered less quickly and T_{CW} will increase).

The QoS decision must be made, for each link, on the receiving node (see next section). A timestamp can thus be transmitted by the transmitting node in a request or data packet's extension to indicate to the receiver the moment when the packet is ready at the MAC layer.

Furthermore, this first approximation depends on the packet's size and it is necessary to carry out a correction to take into account the real data packet's size and not the RREQ packet's length, when those are used for measurement. We thus propose a correction on the d_{AB} duration:

$$d'_{AB} = d_{AB} + \frac{(Data\_length - RREQ\_length)}{Throughput} \tag{4}$$

We can then express the corrected bandwidth:

$$abw' = \frac{Data\_length}{d_{AB} + \dfrac{Data\_length - RREQ\_length}{Throughput}} \tag{5}$$

To preserve the reactive feature of our QoS routing and to avoid intrusive measurements, we choose to not use beforehand, as in the studies [10-11], a temporal window to calculate an average on successive measures with, for example, "hello" packets. In our proposal, the estimate of the available bandwidth is initiated by the first request packets charged to trace the QoS route but uses indifferently all the control or data packets having already crossed the concerned link before the request (passive measurement). A first order smoothing taking into account the previous measurements realized with the request or data packets is then realised. We can thus express the available bandwidth:

$$\widehat{abw}(t) = \alpha.abw(t) + (1-\alpha).abw(t-1) \tag{6}$$

Moreover, to guarantee the validity in time of the measure and to take into account the arrival or the extinction of neighbouring or interfering flows, we propose a mechanism integrated into the routing protocol which informs the source of a reduction of *abw* beyond a preset threshold (see section 4).

Another limitation of our estimate is not specific to the chosen method but more generally related to admission control on a radio medium. Indeed, it is impossible to estimate beforehand the effect, due to the contentions on the medium that a new data flow will have on the available bandwidth measured according to the flow already present in the neighbourhood or interference zones. Thus the admission of a new flow based on the preliminary estimate, the least intrusive possible, will cause more contentions and an inevitable fall of the available bandwidth for all flows. To overcome this issue, one of the usual solutions is to introduce a margin into the admission control to compensate the global reduction in bandwidth (see next section). This solution can be improved with a dynamic re-negotiation when the load in the neighbourhood is modified.

3   QoS Routing Process

The QoS routing objective is to find a route with enough available resources to satisfy a QoS request. Thus, a source having a bandwidth constraint transmits a RREQ packet with QoS bandwidth extension (an optional extension is foreseen by AODV for its main packets RREQ and RREP). This extension indicates the minimum bandwidth that is needed on the whole path between the source and the destination. Before forwarding the RREQ packet, an intermediate node compares its available bandwidth *abw* (note that *abw* is estimated for the concerned link and not for the node receiving the RREQ) to the bandwidth field indicated in the QoS extension (see Fig. 2). If the bandwidth required is not available (i.e. if according to the estimate it does not remain sufficient bandwidth on the concerned link), the packet is discarded and the process stops.

As specified previously, even if the traffic used for measurement does not affect significantly the entire traffic load during the route search, it is necessary to introduce a margin on the comparison to compensate the effect of the new flow on other flows when the route will be established (new contentions will reduce the estimate of the inactivity periods of the channel). For example, for a QoS request of 500kbps, a value of 540kbps for *abw* on one of the links will be insufficient if the selected margin is

10% of the request. This admission control margin has to be tuned according to the traffic characteristics and the density of the network.

In response to a QoS request, the destination sends a RREP packet with the available bandwidth measured on the link which precedes it. Each intermediate node receiving the RREP compares the bandwidth field of the extension with the available bandwidth on the link which precedes it and keeps the minimum between these two values to propagate the RREP on the selected route (Fig. 2). The received value (bandwidth field of RREP) is also recorded in the temporary routing table for the concerned destination. It indicates the minimum available bandwidth for the destination. This entry update allows an intermediate node to answer the next RREQ simply by comparing the minimum bandwidth fields of the table with the value of the transmitted extension (the request does not need to be forwarded toward the destination). This information remains valid as long as the route is valid (in AODV, the lifetime is based on sequence number transmitted in control packets and on the "ACTIVE_ROUTE_TIMEOUT" parameter).

Finally, one should note that the data packets are also used to estimate *abw* and to calculate an average value. This is realized in order to assess the validity in time of link bandwidth measurement. Also, in the case where a particular link is already used by a route, this continuous estimate will serve during new QoS route search involving the concerned link.

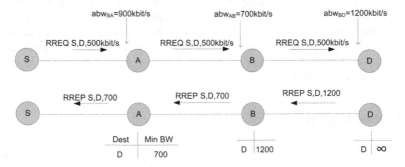

Fig. 2. Example of QoS bandwidth request and response

4 QoS Loss Recovery Mechanism

When because of node movement or new traffic arrival, the available bandwidth decreases on a link of the route, it is possible that the QoS constraint required by the source can not be respected any more. In order to overcome that, a QoS loss recovery mechanism is implemented. The latter uses a predefined QoS Bandwidth Margin (QBM). A route error packet (RERR) is generated when an intermediate node detects a decrease in *abw* that is greater than QBM. As for standard AODV route error mechanism, the RERR packets are sent to all the precursors stored for all the concerned routes. These routes are then erased from routing tables. When the RERR packet reaches the QoS source, it initiates a new route discovery with RREQ packet if the QoS route is still needed.

In the example of figure 3, the link between node A and node B is used for two QoS routes: S1-C-A-B-D1 and S3-A-B-F-D3. When B receives a RREQ packet from A for the QoS route from S1 to D1 (or for any other QoS route discovery involving the link of A towards B), it compares its previous value of *abw* with the new measured one. Then, if it detects a decrease in *abw* that is greater than QBM, the node B send a RERR packet to A which specifies the list of the unreachable destinations (D1 and D3). After reception, node A forwards the RERR packet to all the precursors (C and S3) stored in its routing table for the routes to destinations D1 and D3 and then erases these routes.

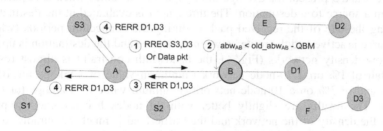

Fig. 3. Example of QoS Bandwidth lost

Note that the *abw* is measured each time a RREQ or a data packet is received by a node, which generally corresponds to a change of the traffic load (new source) or of the network topology (node movement and route failure) producing a possible loss of QoS. Furthermore, if the QBM margin is selected too large, the source node will not be informed of an eventual QoS loss. Conversely, if the margin is too small, useless RERR packets can be generated, causing new RREQ broadcasts. This undesirable control packet transmission induces an undesired overhead, slowing down data packet exchanges, even if the QoS constraint is initially respected. So, an adequate choice of this margin, according to the characteristics of the network, is necessary.

5 Simulation Results

In order to evaluate the performance of our QoS routing protocol, we simulate the proposed mechanisms using NS-2 [16] and its 802.11b extensions.

The radio model allows a signal rate of 11 Mbps, a transmission range of 100m and a detection range of 200m. These ranges correspond to the neighbourhood and interference zones previously described and are close to the typical values obtained for an indoor 802.11b deployment. The number of mobile nodes is set to 10, 20 or 50 nodes giving three simulation sets. These nodes are spread randomly in a 400×400m area network and they move to a random destination every 30s with a speed randomly chosen between 0 and 2m/s (maximal speed for a pedestrian user). Simulations run for 300s.

Traffic sources are CBR (Constant Bit Rate). Their rates correspond to the QoS constraint (i.e. the bandwidth required by the source for its QoS application) and are fixed to 500kbps (122 packets of 512 bytes per second). This value is selected to reach the saturation limits (more than 50 % of the delivered packets are not received) when the number of sources is maximum. Several simulations are realized by varying the number of CBR sources from 10% to 100% of the total number of nodes (a node can integrate several CBR sources). Preliminary measures not represented here made it possible to fix optimal values for the different parameters: 0.2 for the smoothed factor α used for *abw* estimation; 150kbps (30% of the request) for the admission control margin on the bandwidth requests and 50kbit/s (10% of the request) for QBM. For the various network densities, this value of QBM margin corresponds to the best compromise between the reactivity and the overhead and thus gives the best results in terms of global average throughput and number of transmitted packets.

Figure 4 and 5 present the average throughput on all routes when data packets are sent from a source to a destination. The throughput is evaluated at the destination by measuring the rate of the received packets during the connection periods (when the CBR source is active and the route between the source and the destination is up).

For low density networks (Fig. 4), the bandwidth constraint is almost respected with a fall of 1% on a 20-node network (minimum value of 495kbps for 100% of sources) and of 3% on a 10-node network (minimum value of 484kbps for 60% of sources). The results are slightly better with 20 nodes because the compromise between the density of the network and the traffic load is much favourable to search and maintain QoS routes in this case. Without QoS routes, the available bandwidth on the links is not considered and the delay between two packet's receptions on the selected route can increase significantly, thus reducing the average global throughput (from 500kbps to 420kbps for 20 nodes and 100% of sources).

For a high density network of 50 nodes, the improvement is still better (Fig. 5). With QoS, the throughput is always higher than 470kbps (6% less compared to the constraint). When the number of QoS sources reaches 100%, the average global throughput slightly decrease (440kbps) because of the transitional periods corresponding to the route reorganizations are more frequent. Without QoS, the throughput is not controlled at all and it decreases very quickly: less than 250kbps (50% of the constraint) as soon as the number of sources becomes higher than 20%.

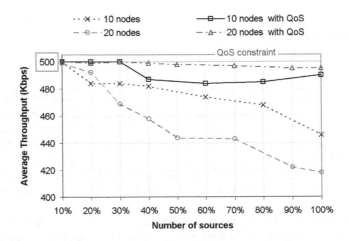

Fig. 4. Average Throughput / Number of sources (10 and 20 nodes)

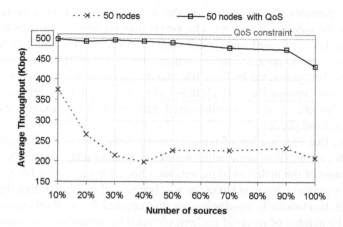

Fig. 5. Average Throughput / Number of sources (50 nodes)

Figure 6 shows the evolution of the number of packets which are dropped along the way when the number of sources increases. The y-axis thus represents the ratio between the number of packet dropped by an intermediate node (whatever the reason: mobility or contention) and the number of data packet sent by the source when a route is supposed to exist. In all the cases, the results are clearly better with QoS: less than 12% of dropped packets whatever the density of the network. These important improvements are due principally to two factors: the rejection of the unreliable routes during the QoS route setup procedure and the QoS loss notifications mechanism which avoid sending packet when a high contention risk is foreseen. Note that without QoS, the number of dropped packets can be very important (up to 86% with 50 nodes), confirming that the accepted routes are not reliable.

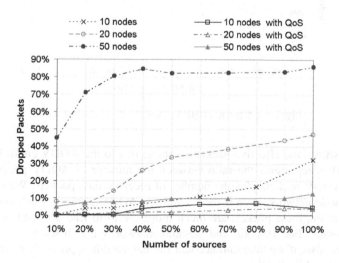

Fig. 6. Packets Dropped on the way / Number of sources

Figure 7 presents the evolution of the total number of undelivered packets according to the number of sources. The y-axis thus represents the ratio between the number of packets generated by the CBR sources which never reach their destinations (whatever the reason: no route at the source, mobility or contention) and the total number of packets generated by the CBR sources. Let us note that this latter is about identical with or without QoS, which thus makes it possible to compare, with theses curves, the absolute number of undelivered packets or, by difference, the absolute number of received packets.

We notice that the number of undelivered packets is not very different with or without QoS routing (it is even better with QoS in a 50-node network). This result shows that most of the undelivered packets are rejected at the source because no route is possible or even are dropped along the way because of mobility rather than because the available bandwidth is insufficient on the crossed links. Furthermore, we can notice that the number of received packets, obtained by subtraction, is also quite close with QoS but for all these received packets, the QoS condition is met (remember that without QoS the average throughput can be less than 250kbps). This show that the QoS routes are not obtained to the detriment of the number of delivered packets.

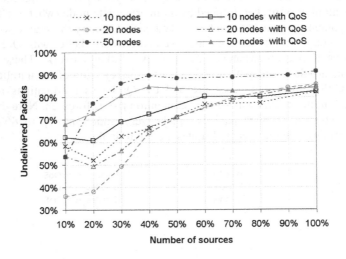

Fig. 7. Undelivered CBR Packets / Number of sources

Figure 8 shows the effective packet overhead due to the AODV control messages. The y-axis thus represents the ratio between the number of AODV packets and the number of AODV packets plus the number of received data packets. We can observe that for 50-node network, even if there are more AODV packets which circulate with the QoS solution, these packets are much more useful to find a reliable route and to avoid the losses. For 10-node and 20-node networks the difference is very weak showing here also, if we take into account the bandwidth improvements (Fig. 4), that the QoS overhead is effective.

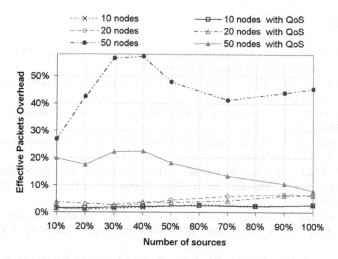

Fig. 8. Effective Overhead / Number of sources

6 Conclusion

In this paper, we proposed and evaluated a QoS routing solution for ad hoc networks based on link available bandwidth monitoring. This solution uses standard protocols (AODV for the routing process and 802.11 at the link layer) and is intended for small and dynamic ad hoc networks in which an important reactivity is necessary. The QoS routes are traced node by node and the proposed routing algorithm uses only extensions of the AODV request (RREQ) and reply (RREP) packets. The bandwidth measurement is initiated on RREQ arrivals in a node (these times correspond to a network state change: arrival of a new flow or route recovery following a node movement) to avoid increasing the overhead unnecessarily. Moreover, a QoS loss recovery mechanism, also based on the existing extensions, is used to take into account the variations of the available bandwidth due to the dynamic nature of the ad hoc network and its traffic.

The proposed QoS routing with QoS loss recovery gives very satisfying results. The required bandwidth on the QoS routes is obtained with less than 6% of fall in almost all the cases. Comparatively, this bandwidth that the source nodes seek to maintain thanks to the QoS routing can be divided by two in the absence of this one. In addition, these improvements are not obtained to the detriment of the number of routed packets since in all the cases, the relative rate of dropped packets is lower with the QoS solution. Lastly, the effective overhead is less important with the QoS constraint whatever the density of the network.

As for future works, we are targeting the issue of the dynamic optimization of QDM and the admission control margin. A first idea will be to analytically derive a relationship between these margins and both the number of flows crossing each node and/or their bandwidth/delay requests.

References

1. IEEE 802.11 WG, Part 11: Wireless LAN Medium Access Control (MAC) and Physical Layer (PHY) specifications, Standard, IEEE, 1999.
2. IEEE 802.11g WG, Part 11-Amendment 4: Further Higher-Speed Physical Layer Extension in the 2.4 GHz Band, 2003
3. C. E. Perkins, E. M. Royer, and S. R. Das, "Ad hoc on-demand distance vector routing," Internet Draft, 2002.
4. C. E. Perkins, E. M. Royer "Quality of service for ad hoc on-demand distance vector routing". IETF Internet Draft.
5. J. Hsu, S. Bhatia, M. Takai, R. Bagrodia, M. Acriche, "Performance of Mobile Ad Hoc Networking Routing Protocols in Realistic Scenarios," MILCOM '03 (Boston, MA, October 2003).
6. A. Agarwal and W. Wang, "Statistical Analysis of the Impact of Routing in MANET Based on Real-Time Measurements," ICCCN'05, San Diego , California , USA , October, 2005.
7. L. Chen, W. B. Heinzelman "QoS-Aware Routing Based on Bandwidth Estimation for Mobile Ad Hoc Networks" IEEE Journal on Selected Areas in Communications, March 2005.
8. C. Chaudet, I. Guérin Lassous - "BRuIT - Bandwidth Reservation under InTerferences influence" - In Proceedings of European Wireless 2002 (EW2002) - February 2002 - Florence, Italy.
9. K. Xu, K. Tang, R. Bagrodia, M. Gerla, and M. Bereschinsky, "Adaptive Bandwidth Management and QoS provisioning in Large Scale Ad Hoc Networks," in Proceedings of MilCom. Boston, Massachusetts,USA: IEEE, Oct. 2003.
10. M Kazantzidis, "End-to-end versus explicit feedback measurement in 802.11 networks," Technical Report N° 010034 UCLA Computer Science WAM Lab,. 2001.
11. S. H. Shah, Kai Chen, Klara Nahrstedt, "Dynamic Bandwidth Management for Single-hop Ad Hoc Wireless Networks", ACM/Kluwer Mobile Networks and Applications (MONET) Journal, Special Issue on Algorithmic Solutions for Wireless, Mobile, Ad Hoc and Sensor Networks, vol. 10, num. 1, 2005.
12. The Network Simulator - NS-2: http://www.isi.edu/nsnam/ns/

An Efficient and Robust Service Discovery Protocol for Dynamic MANETs

Mohammad Nazeeruddin, Gerard Parr, and Bryan Scotney

School of Computing and Information Engineering
University of Ulster, Coleraine, Northern Ireland
{nazeer, gparr, bryan}@infc.ulst.ac.uk

Abstract. Automatic service discovery is essential for the usability of
self-configuring Mobile Ad Hoc Networks (MANETs). Existing service
discovery solutions are either directory based or directory-less. The directory based approaches are fast and reliable but have the complexity
of maintaining directories in the dynamic MANET environment. On the
other hand, the directory-less approaches use broadcasts which are very
expensive in MANETs. This paper proposes and evaluates a new service discovery protocol called MANET Service Location and Discovery
(MSLD), which is integrated with a stateful auto-configuration protocol
called DHAPM. Because of this integration, MSLD reuses the robust
directory structure of DHAPM, which allows MSLD to acquire the advantages of the directory based service discovery protocols without any
additional complexity of directory maintenance. The performance evaluation shows that MSLD has low service discovery latency and high
service availability with minimum communication overhead.

1 Introduction

The tremendous growth of portable handheld devices (e.g. cell phones, iPAQs,
PDAs) with basic networking capabilities (Bluetooth, GPRS, 802.11) has opened
new research directions in the area of networking [1]. Mobile Ad hoc Networks
(MANETs) provide the necessary capabilities to form a rapid, self-organizing,
infrastructureless network among mobile wireless peers without any prior planning. One of the key reasons behind the formation of a MANET is to effectively
utilize the services provided by the individual peers (nodes or hosts) in the network. Nevertheless, to effectively exploit the services provided by the peers, a
simple robust mechanism for discovering services is needed. Service location or
discovery protocols offer a convenient and effective framework for the networked
hosts to advertise their services to other hosts and access the information about
the existence, location and configuration details of the other networked services [2]. Services on a device include accessible software components, hardware
components and data which other devices may need. For instance, in military
communications data from several heterogenous devices, offered by different services, should be integrated to discover meaningful trends [1].

Because of several inherent limitations of mobile devices and wireless
networks, enabling service discovery is a challenging task. Thus, the MANET

A. Helmy et al. (Eds.): MMNS 2006, LNCS 4267, pp. 49–60, 2006.

Service Discovery (SD) protocols should have the following features. (i) Low overhead: mobile nodes are limited-resource devices and hence MANET protocols should use minimal device resources. (ii) Robustness: MANET protocols should tolerate intermittent network connectivity (which arises because of node mobility) and device failures. (iii) Versatility: any network enabled device can be part of a MANET. So the SD protocols should be able to work with a wide variety of devices. (iv) Scalability: MANETs have diverse application ranging from small personal networks to diverse military networks. Thus, the MANET protocols should work seamlessly in different networks.

In this paper we propose a service discovery protocol for MANETs and show its integration with DHAPM [11] [12], a dynamic host auto-configuration protocol for MANETs. The proposed service discovery protocol is lightweight, robust and can adapt dynamically to the network conditions.

2   Related Work

Theoretically the service discovery problem can be solved by using either a directory based (centralized) approach or a directory-less (distributed) approach.

In a directory based approach, a directory of the available services in the network is maintained by a single host or a group of hosts. Service agents (Servers) register their services along with the service access details in the directory and clients query the directory to obtain the service descriptions.

In the directory-less approaches, as there are no centralized directories, the service messages should to be distributed to all nodes. Depending on distributed message type, the service discovery can be done in two modes. (i) Push mode: servers broadcast (or multicast) frequent service advertisements to all hosts and clients passively cache the services and select the service which is of interest. (ii) Pull mode: clients broadcast service requests and servers that offer the requested service reply to the request.

The traditional service discovery protocols designed for wired networks (like SLP, Jini, UDDI), do not account for the latency and packet loss issues associated with the several intermediate wireless links connecting multi-hop nodes and intermediate node mobility [8]. Moreover, the traditional protocols do not take account of the device heterogeneity and resource constraints. In general, there has been an agreement among researchers that the traditional protocols cannot be used in MANETs [3, 8].

Some new SD protocols for MANETs were developed specifically for a physical layer technology, like Bluetooth's service discovery protocol and IrDA's information access protocols [3]. Some other protocols were developed as part of high-level distributed application technology (e.g. SD mechanisms defined in JXTA and OSGi) and other approaches (e.g. INS and one.world [3]) developed as part of complex pervasive computing architecture. Though these approaches are efficient in the specific environment for which they were developed, they are not suitable for general heterogenous MANETs.

The specific ties to a particular environment can be eliminated by designing SD solutions at the application layer of the protocol stack. As MANETs are infrastructure-less networks, the obvious choice for the SD protocols is a directory-less architecture and hence several new directory-less protocols (Konark, DEAPspace, PDP [3], etc) have been proposed for MANETs. The common problem in the directory-less (pull-based or push-based) approaches is that they employ broadcasting or multicasting both of which add substantial communication overhead [7]. Also, these approaches suffer from scalability problems in larger networks [13].

The communication overhead can be minimized by using distributed service directories. In [7] a service discovery solution is proposed based on the service directories which are deployed dynamically. It was shown that this type of solution minimizes the generated traffic. However, the directory based approaches have an additional overhead of selecting the directory servers and maintaining the directories.

The proposed MANET Service Location and Discovery (MSLD) protocol also employs distributed directories to achieve robustness against directory failures. MSLD is integrated with DHAPM and uses the DHAPM address agents (AAs) to perform the role of service directories. Thus, MSLD does not have any additional overhead of selection and maintenance of directories while having low communication overhead and the robustness offered by the directory-based approaches.

3   Integration with DHAPM

Dynamic Host Auto-configuration Protocol for MANETs (DHAPM) [12] is a stateful host auto-configuration protocol based on multiple dynamically selected Address Agents (AAs), which deliver the important functions of auto-configuration. Each Address Agent (AA) manages a disjoint block of sequential IP addresses which it can assign to a new node without consulting any other AA in the MANET [10]. Each AA maintains two tables - an address table and an agent table. The Agent table keeps the information about all other AAs in the MANET. The Address table keeps a record of all the assigned IP addresses and the details of nodes to which they are assigned. Node details include hardware address, node status, and a boolean bit [12]. AAs periodically synchronize with the other AAs by sending incremental address table updates.

DHAPM has robust mechanisms for initiating and maintaining the AAs. The AAs in DHAPM can be easily extended to support service discovery tasks. The following are the required amendments to DHAPM for providing Service Discovery related tasks:

- Service directory: A service directory holds a list of available services including the details of the service providers and service attributes. DHAPM already maintains hostnames and IP addresses in the address table for auto-configuration and name resolution[1]. So the same table can be supplemented with an extra "services" column as shown in Table 1. Since the address table is sorted with respect to IP addresses, the whole table should be searched to find all the matching services for a given requested service type, which is

---

[1] Name resolution related tasks are provided by a separate protocol called MNS.

inefficient. For this reason a separate sorted table called "service directory" is maintained. The service directory is derived from the address table and sorted with respect to service names (types) for efficient service lookups.

– A mechanism for querying/administering database: Once a service directory is available, DHAPM should provide a mechanism to query and administer this database. To carry out this task, a new module called MANET Service Location and Discovery (MSLD) is added to the DHAPM protocol. MSLD is described in Section 4.

Table 1. Modified Address Table Accommodating Service Information

IP	MAC	Name	Service(s)	Status	Suitable for AA	ALT
1	Mac1	PC1	srv1@attr1=val1;attr2=val2	0	TRUE	3600
2	Mac2	PDA2	srv1@attr1=val1;attr2=val2	4	TRUE	600
3	Mac3	PC3		1	FALSE	200
.....	.....	.....		...	.....	...
253	Mac253	LT	srv10@attr1=val1, srv13@attr1=val1	0	TRUE	500
254	Mac254	PDA	srv4attr1=val1;attr2=val2;attr3=val3	1	TRUE	400

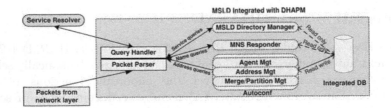

Fig. 1. MSLD Integrated with DHAPM

The integration of SD protocol in DHAPM brings the following benefits:

– MSLD has no additional overhead of Directory Manager (DM) selection and maintenance while having the benefits of a distributed directory based approach.
– MSLD service directory can use the existing information from the address table instead of re-requesting from the services.
– Basic service registration and service deregistration can be done during the address allocation and address relinquishment process respectively.
– MSLD module runs as a separate thread of DHAPM process instead of an independent process which saves the system resources.

The integrated architecture of DHAPM with service discovery module is shown in Fig. 1. DHAPM query handler receives all queries including service discovery and name resolution queries. Depending on the type of a query, the DHAPM query handler forwards it to an appropriate module. The integrated database (address table, agent table and service directory) is managed by autoconf module and the other modules have a Read-Only (RO) access to the database.

4   MANET Service Location and Discovery (MSLD)

MSLD has three components: Service Resolver (SR), Service Manager (SM) and Directory Manager (DM).

Service Resolver (SR): Service resolver is a software routine which works on behalf of the user applications to acquire information about the available services with their attributes and configuration details. SR retrieves the service information in two modes depending on whether a DM is accessible or not: (i) Unicast mode (U_C mode): SR operates in this mode when it has information about one or more DMs. In this mode service discovery requests (SDQs) are unicasted directly to a DM and DM unicasts the service discovery replies to the requesting SR. By default SR operates in this mode. (ii) Broadcast mode (B_C mode): SR temporarily switches to this mode when it cannot communicate with any DM. In B_C mode SR broadcasts SDQs in the MANETs. If any device is offering the requested service then it unicasts a service discovery reply (SDY) with the service details to the requesting SR. Moreover, DMs also reply to the broadcasted requests indicating their availability for any matching entries from their service directory. If SR receives a reply from a DM, then it switches back to U_C mode.

SR caches the retrieved service details for a certain lifetime called Cache Life Time (CLT). If the same request is issued again before the elapse of CLT seconds, then SR replies immediately from the cache instead of sending a service discovery request (SDQ) to a DM. Once the CLT of a service expires, the service information becomes obsolete and is deleted from cache.

Service Manager (SM): The service manager manages all services offered by a device. It serves two purposes: (i) communicates with a DM on behalf of all services[2] available on the devices; the services register their information (service attributes and configuration) directly with the local SM instead of independently registering with a DM; (ii) replies to the broadcasted SDQs, which is essential for the smooth functioning of service discovery protocol in the absence of DMs.

Using a SM as an intermediate broker not only optimizes the communication overhead but also relieves service developers from the burden of writing additional code to interact with DMs.

Directory Manager (DM): A MSLD directory manager also performs two tasks: (i) collects and maintains the information about all the services available in the MANET in the form of a service directory; (ii) resolves the received service discovery requests using the service directory.

The DM receives SDQs from either a local SR or remote SRs running on the other hosts. Whenever DM receives a SDQ, it just checks its service directory for the requested service type. If it finds one or more entries then it returns the matching set. Otherwise it returns a null set with an error code. When the DM is bombarded with several simultaneous SDQs, it buffers the SDQs and serves each request using a First In First Out (FIFO) policy. If the buffer limit, which

---

[2] Services and service agents (SAs) are interchangeably used in this document.

is determined based on the individual DM capacity, is exceeded then the DM drops the new requests.

4.1 Service Management and Lookup

This subsection describes the details about the MSLD service directory management and service lookup.

Service Naming Convention: Any networked service may be encoded in a service Uniform Resource Locator (URL). MSLD employs compact service URLs tailored for MANETs using similar syntax and semantics defined in [9]. Having a compact service URL not only optimizes the storage requirements of SD but also saves the energy needed while transmitting and receiving service URLs. A service URL of MSLD protocol can be encoded in the following form:

$$< srvtype >: // < addrspec >; < attrlist > .$$

The service type ($< srvtype >$) describes the type of the service. It contains an abstract name (e.g. printer) and is sometimes supported by a concrete type (printer:lpr) [9]. An address specification ($< addrspec >$) in the service URL is the hostname or IP address of the service provider. An attribute list ($< attrlist >$) is a string containing all the attributes of a service separated by semicolons. Each attribute and value are expressed as "$attribute = value$". CLT is also added as an attribute. The following are some example service URLs:

$$printer://host1.edu;duplex=TRUE$$
$$videochat://host2;authentication=KERBEROSV4;speed=5fps$$

It should be noted, however, that MSLD is not tied to any particular service description syntax, and any other service description formats can be used instead of the proposed format.

Service Registration: Whenever a service is started on a device the corresponding service agent (SA) registers itself with the local Service Manager (SM). The SM coordinates the service registrations with a DM.

The initial service registration is done as part of the new address request process. Whenever a new node requests an address, it also informs the AA[3] about the services hosted by it (including the $< srvtype >$ and optional $< attrlist >$). The address specification part of the service URL ($< addrspec >$) is omitted because AA can fill that part itself. AA acknowledges the service registration when it actually assigns an address to the new node. Each registered service has a lifetime, which defines how long the service entry will be valid. If the SA does not specify a lifetime then the address lease time (ALT) of the host will be used as its default lifetime. Services should be renewed (or re-registered) before the service lifetime expires.

A node can also register its services any time after the initial registration using a *SrvReg* message and DM acknowledges the registration using a *SrvRegAck*

---

[3] It should be noted that DM is an integral part of AA. For the sake of clarity the term AA is used whenever referring to the specific autoconf tasks and the term DM is used when referring to the service discovery tasks.

message. In addition, devices may need to change the registered service infor-mation (service attributes) from time to time. This can be done by sending the updated service information to its DM using a *SrvUpd* message. DM replaces the existing service details with the new details and acknowledge the updates using a *SrvUpdAck* message.

Service Lookup: Whenever a service discovery request (SDQ) is received from user application, Service Resolver (SR) first checks whether the requested service query has some matching entries in the local MSLD cache. If some matches are found with valid CLT then the SR immediately returns the matching entries to the application with an indication that the matching set is from the local cache. If the application is not interested in the cached entries, it may re-request SR to query a DM for the updated service information.

If no matches are found in the local cache or if the application requested specifically to query a DM, then the SR constructs a SDQ message, forwards the SDQ message to its DM (AA), and starts a SDQ Timer. If a service discovery reply (SDY) is received from the DM, then SR extracts the results from the reply message and forwards them to the requested application. In case the received SDY contains a null set, SR returns an appropriate message to the application. Conversely, if no response is received from the DM and the timer expires then SR re-sends the request to its DM until a response is received or the number of retries exceeds r_{min}. If there is no response even after r_{min} retries then the resolver forwards the request to the DM_{pid}[4] (AA_{pid}) with a probable departure notification of its DM (AA) and restarts the SDQ Timer. If the DM_{pid} also does not respond then the resolver switches to B_C mode (i.e. directory-less mode).

In B_C mode SR broadcasts SDQs so that service managers (SMs) can directly reply to the SDQs. When the SR first switches from U_C mode to B_C mode, it broadcasts the old SDQ only one time and for the subsequent queries SR retries a request for r_{min} times. If the broadcasts also fail to elicit any responses then the SR returns an error message to the application and terminates the request process. If SR receives a reply from a DM at any point in time then it switches back to U_C mode to reduce the excessive overhead due to the broadcasts.

Service Deregistration: When a node intends to leave the MANET it sends an address relinquish message (*AddrRelReq*) to its AA. After receiving this message from the departing node, the AA marks the nodes's record, including any services offered by the departing node, as invalid. In the case of abrupt node departures, records will be automatically deleted after the address lease time (ALT) of the host expires. This task is handled by autoconf module, so there is no additional overhead on MNS. Also, a service agent can deregister any of its services at any time using a *SrvDeReg* message.

5   Mathematical Analysis

In this section the communication overhead of different alternative approaches is compared with MSLD protocol.

---

[4] AA_{pid} is the node which generated MANET PID. For details refer to [10].

Message overhead of the directory-based approaches: An ideal directory based approach has a directory and all devices in the network register their services in this directory. The directory announces its presence periodically by flooding announcement messages. Clients unicast the SDQs to the directory and the directory unicasts the replies to the clients. Assume N_χ is the average number of SDQs, N_{reg} is the average number of service registrations, and N_{cng} is the average number of service updates during the each directory announcement period (T_{ad}). Then the message overhead incurred during T_{ad} is:

$$B_C + 2 * N_\chi * U_C + 2 * N_{reg} * U_C + 2 * N_{cng} * U_C \tag{1}$$

where B_C and U_C is the message overhead of a broadcast and a unicast message respectively.

Message overhead of the directory-less approaches: In the push-based directory-less approaches the service agents periodically announce their services and clients cache the announced service information. If N_{sp} is the average number of service providers then the message overhead incurred during T_{ad} is:

$$N_{sp} * B_C \tag{2}$$

In the pull-based approaches clients flood the SDQs and service agents unicast the replies. If α is the average number of service agents providing the similar services then the message overhead incurred during T_{ad} is:

$$N_\chi(B_C + \alpha * U_C) \tag{3}$$

Message overhead of the MSLD: MSLD maintains a distributed directory, which is periodically synchronized to achieve robustness against node failures. As MSLD is integrated with DHAPM, directory synchronization is done automatically as part of AA synchronization. Only the AA synchronization packet will have extra *service* data. Moreover, the initial service registration is done as part of the address request process. Only the registration of services which started after the initial service registration and service updates will be sent separately. If ζ is the directory synchronization overhead and N_{reg}^- is the average number of late service registrations during the period (T_{ad}) then the message overhead incurred during T_{ad} is:

$$\zeta + 2 * N_\chi * U_C + 2 * N_{reg}^- * U_C + 2 * N_{cng} * U_C \tag{4}$$

Comparing Eqs. 1 and 4, as $\zeta \ll B_C$ and $N_{reg}^- <= N_{reg}$, it is evident that the overhead of MSLD (integrated with DHAPM) is always less than directory based approaches.

In the case of directory-less approaches, both push and pull based approaches employ flooding for SD which is very costly in MANETs. Although the overhead of flooding can be minimized by using the restricted broadcast (restricting broadcast to few hops), still its overhead is much greater than a unicast message [6]. Thus, comparing Eqs. 2 and 3 with Eq. 4, it is evident that MSLD overhead is less than both the push and pull based directory-less approaches.

Comparison with routing-based approaches: It has been argued in the literature that the service discovery message overhead can be reduced by piggybacking the service discovery messages with the underlying reactive protocol messages. It was demonstrated in [5] through extensive simulations that pull-based approaches minimize the overall broadcasts when SD messages are integrated with the reactive routing protocol messages. This sub-section analyzes the overhead of MSLD and directory-less approaches when such an optimization technique has been supported by the underlying routing protocol.

As the forward and the reverse routes to the service provider are established during the service discovery process itself, the overall routing and SD message overhead per SDQ will remain the same in directory-less approaches and is given by $B_C + \alpha * U_C$. However, in MSLD the overall overhead will be different depending on whether a route is already available or not. The following three cases are possible in MSLD:

- Case 1: when the route to any DM is not known: in this case, the client sends the route discovery broadcast integrating SDQs. The client receives unicast replies containing service information from the DMs and service managers. So the overall overhead is similar to the pull-based approach ($B_C + \alpha * U_C$).
- Case 2: when the route to a DM is known: in this case, the client unicasts a SDQ to the DM and the DM unicasts the reply. Then the client selects a service provider (SP) and contacts it. If a route to the SP is already known then the overall overhead per SDQ is: $\zeta/N_\chi + 2 * U_C + 2 * N_{cng} * U_C/N_\chi \approx 2 * U_C$ (for large values of N_χ). This implies that MSLD still has lower overhead.
- Case 3: when the route to DM is known and SP route is unknown: in this case, the additional route discovery overhead excluding SP contact overhead ($B_C - U_C$) should be added to the overhead in case 2. Therefore, the overall overhead is $B_C + U_C$. Comparing with the pull-based system overhead (Eq. 3), the overhead of MSLD is lower only if $\alpha > 1$. Otherwise the pull-based approach has better performance. Nevertheless, when the requested service is not available in the network (i.e. $\alpha = 0$), MSLD needs only $2 * U_C$ and pull-based approach needs $r * B_C$. This means MSLD is more efficient.

From the above analysis, it is evident that the pull-based approaches integrated with a routing protocol may have lower overall communication overhead only in case 3 when $0 < \alpha < 1$. In all the other cases MSLD has better or similar performance than the pull-based approaches. In short, MSLD is still an efficient solution even when the service discovery messages are piggybacked on the routing messages.

6 Simulation Based Analysis

From the mathematical analysis it is evident that MSLD has lower communication overhead than the alternative approaches. In this section, we evaluate MSLD performance based on two other performance metrics (service discovery

latency and service availability) using the ns-2 simulator. All simulations were
executed for a duration of 300 seconds. We considered 50 mobile terminals with
transmission range of 250m randomly distributed over the simulation surface. To
simulate dynamic MANETs we used the random way point mobility model with
max speed of 4m/s, min speed of 2m/s and zero pause time. In the simulations,
all the service discovery requests are uniformly distributed over 150 seconds. All
the nodes in the MANET generate an equal number of requests.

The two service discovery architectures simulated are MSLD and directory-
less (distributed) architectures. The primary reasons behind the simulation of
directory-less architecture are (i) it is an obvious alternative to MSLD; (ii) MSLD
switches to directory-less mode in the absence of DMs.

Service Discovery Time (Latency): The service discovery latency of a protocol is the time needed to resolve a service type to a valid service binding and
contacting the service agent successfully through the obtained service binding.

To evaluate the latency, we have executed several simulation scenarios where
clients are 1-6 hops away from the DM and server node. For each hop the mean
(μ) and standard deviation (σ) of service discovery resolution time, service agent
contact time and total binding time were calculated. The obtained results are
summarized in Tables 2 (a) and (b) for MSLD and directory-less protocols respectively. Notice that latency increases with the increase in the distance (number of hops). This is because of the accumulated packet forwarding delays at the
intermediate nodes. However in the case of MSLD, the total latency is less than
50ms even if the client is 6 hops (around 1500m) away from both DM and SP
which is negligible. On the other hand, the directory-less approaches have higher
latency because of the broadcasting delays.

Table 2. Latency (ms) Vs Hops

	(a) MSLD						(b) Directory-less				
Hops	μ_{sdy}	σ_{sdy}	μ_{sp}	σ_{sp}	μ_{total} σ_{total}	Hops	μ_{sdy}	(σ_{sdy})	μ_{sp}	(σ_{sp})	μ_{total} (σ_{total})
1	3.26	(0.25)	3.43	(0.22)	6.69 (0.34)	1	8.24	(3.50)	4.54	(2.25)	12.78 (4.31)
2	6.73	(0.37)	6.87	(0.35)	13.61 (0.52)	2	14.97	(4.22)	7.73	(0.79)	22.71 (4.10)
3	10.21	(0.44)	10.43	(0.45)	20.64 (0.63)	3	22.74	(5.14)	11.40	(1.30)	34.13 (5.33)
4	13.73	(0.54)	13.95	(0.52)	27.68 (0.73)	4	30.34	(5.93)	14.33	(1.82)	44.67 (6.41)
5	17.27	(0.58)	17.44	(0.60)	34.71 (0.83)	5	37.72	(6.34)	17.49	(0.63)	55.22 (6.42)
6	20.74	(0.63)	20.96	(0.62)	41.70 (0.87)	6	45.38	(7.25)	22.24	(2.69)	67.62 (8.30)

Service Availability: Service availability is one of the important performance
metrics to measure the service discovery protocol performance. The service availability of the protocol can be defined as the ratio of successful service bindings
to the total number of requests generated. A successful service binding means
the successful resolution of the service type and a successful contact with the
server [5].

Table 3 summarizes the failed service discovery queries (F_{sdqs}), failed service
bindings ($F_{bindings}$) and service availability of both MSLD and directory-less

approaches for different network areas and SDQ rates. Observe from the table that the total number of failures ($F_{sdqs} + F_{bindings}$) increases with the increase in the network size, resulting in a lower service availability. This is because as the network size increases the node density decreases and hence there is an increased chance of node isolation/separation from the MANET.

An important point to be noted from this is that MSLD has minimal (almost zero) service type resolution failures, which shows the significance of maintaining a distributed service directory. However, service binding failures ($F_{bindings}$) are much higher in MSLD. This is because for a successful service type resolution, the client just needs one accessible DM; but the service binding with the server may fail because of node mobility (both server and client mobility) during the time of service registration and at the moment the client application seeks to contact the service agent. The probability of such failures increases when the node density decreases.

On the other hand, in the directory-less approaches the server itself resolves the service discovery request, so there is a low probability that the server will be inaccessible to the client after a successful service type resolution and hence there are fewer $F_{bindings}$. Nevertheless, the initial service resolution is more difficult in the directory-less approaches as it merely depends upon the availability/accessibilty of the server providing such a service.

Although there are differences in F_{sdqs} and $F_{bindings}$ values, on average the overall service availability is higher in MSLD than in directory-less approaches.

Table 3. Service Availability Vs Network Size

Network Area (m^2)	$\Sigma SDQs$ (SDQ Rate/sec)	MSLD			Directory-less		
		F_{sdq}	$F_{bindings}$	SA	F_{sdq}	$F_{bindings}$	SA
250x250	2000 (13.33)	0	0	1.0000	0	0	1.0000
	4000 (26.67)	0	0	1.0000	1	0	0.9998
500x500	2000 (13.33)	1	43	0.9780	27	24	0.9745
	4000 (26.67)	2	66	0.9830	56	15	0.9822
750x750	2000 (13.33)	3	99	0.9490	57	25	0.9590
	4000 (26.67)	2	169	0.9573	458	177	0.8413
1000x1000	2000 (13.33)	7	143	0.9250	76	66	0.9290
	4000 (26.67)	4	238	0.9395	513	189	0.8245

7 Conclusion

The existing service discovery protocols including the recent solutions tend to add substantial overhead (either for maintaining directories or for flooding the messages) which reduce their applicability in resource-constrained MANETs. The solution proposed in this paper - MSLD, overcomes this limitation by employing distributed service directories. The proposed protocol can also be seamlessly integrated with any directory-based protocols to reduce directory maintenance overhead. An example scenario showing the integration of MSLD with DHAPM (a robust auto-configuration protocol previously proposed and evaluated by us) was described in this paper and the efficiency of this integrated protocol was investigated through mathematical analysis and simulations.

References

1. D. Chakraborty, A. Joshi and Y. Yesha, "Integrating service discovery with routing and session management for ad-hoc networks," Ad Hoc Networks, Vol 4, Issue 2, March 2006, Pages 204-224.
2. E. Guttman, C. Perkins, J. Veizades, and M. Day, "Service Location Protocol Version 2", RFC 2608, July 1997.
3. C. Campo, C. Garcia-Rubio, A. Marin Lopez and F. Almenarez, "PDP: A lightweight discovery protocol for local-scope interactions in wireless ad hoc networks," Computer Networks, In Press, Corrected Proof, Available online 19 January 2006.
4. C. E. Perkins, E. M. Royer, and S. R. Das, "Ad Hoc On-Demand Distance Vector (AODV) Routing," IETF Mobile Ad Hoc Networks Working Group, IETF RFC 3561, July 2003.
5. P. E. Engelstad, Y. Zheng, "Evaluation of Service Discovery Architectures for Mobile Ad Hoc Networks," the Second Annual Conference on Wireless On-demand Network Systems and Services (WONS05) (2005)
6. L. M. Feeney, M. Nilsson, Investigating the energy consumption of a wireless network interface in an ad hoc networking environment, IEEE INFOCOM, Volume 3, April 2001 Pages:1548 - 1557.
7. F. Sailhan, V. Issarny, "Scalable Service Discovery for MANET," Third IEEE International Conference on Pervasive Computing and Communications, 2005. PerCom 2005, Page(s):235 - 244.
8. C. K. Toh, "Ad Hoc Mobile Wireless Networks. Protocols and Systems", Prentice Hall PTR, New Jersey, 2002, Chapter 12, Pages: 231-242.
9. E. Guttman, C. Perkins, and J. Kempf, "Service Templates and service: Schemes", RFC 2609, June 1999.
10. M. Nazeeruddin, G. Parr and B. Scotney, "Fault-tolerant Dynamic Host Auto-configuration Protocol for Heterogeneous MANETs", Proc. of 14th IST Mobile & Wireless Summit, Dresden, Germany, 19-23 June 2005.
11. M. Nazeeruddin, G. Parr and B. Scotney, "A New Stateful Host Auto-configuration Protocol for Digital Battlefield MANETs", Proc. of IEEE MILCOM, Atlantic city, NJ, 17-20 Oct 2005.
12. M. Nazeeruddin, G. Parr and B. Scotney, "DHAPM: A New Host Auto-configuration Protocol for Highly Dynamic MANETs", Accepted for publication in Journal of Network and Systems Management, Oct 2006.
13. J. Tyan and Q. Mahmoud, "A Comprehensive Service Discovery Solution for Mobile Ad Hoc Networks," Journal of Mobile Networks and Applications, 10, 2005, pages: 423-434.

Modeling Energy Consumption in Error-Prone IEEE 802.11-Based Wireless Ad-Hoc Networks

Tsung-Han Lee, Alan Marshall, and Bosheng Zhou

School of Electrical and Electronic Engineering, Queen's University of Belfast, UK
{th.lee, a.marshall, b.zhou}@ee.qub.ac.uk

Abstract. In the IEEE 802.11 MAC layer protocol, there are different trade-off points between the number of nodes competing for the medium and the network capacity provided to them. There is also a trade-off between the wireless channel condition during the transmission period and the energy consumption of the nodes. Current approaches at modeling energy consumption in 802.11-based networks do not consider the influence of the channel condition on all types of frames (control and data) in the WLAN. Nor do they consider the effect on the different MAC and PHY schemes that can occur in 802.11 networks. In this paper, we investigate energy consumption corresponding to the number of competing nodes in IEEE 802.11's MAC and PHY layers in error-prone wireless channel conditions, and present a new energy consumption model. Analysis of the power consumed by each type of MAC and PHY over different bit error rates shows that the parameters in these layers play a critical role in determining the overall energy consumption of the ad-hoc network. The goal of this research is not only to compare the energy consumption using exact formulae in saturated IEEE 802.11-based DCF networks under varying numbers of competing nodes, but also, as the results show, to demonstrate that channel errors have a significant impact on the energy consumption.

Keywords: IEEE 802.11, DCF, Ad-Hoc Energy consumption, error-prone.

1 Introduction

Ad-hoc wireless networks are currently receiving a significant amount of interest. Some of this interest may be attributed to the distributed nature of IEEE 802.11's DCF, which allows for instant deployment and routing of packets around nodes in multi-hop ad-hoc wireless networks. However, in the IEEE 802.11 DCF, the wireless channel needs to be shared efficiently among contending nodes, and considerable research efforts are being dedicated to improving the energy consumption in these networks. In [1] the power consumption of the wireless network interface card was measured when used by different end-user devices. In [2] an analytical model to predict energy consumption in saturated IEEE 802.11 single-hop networks under ideal channel conditions is presented. This work focused on the energy consumption in a condition which assumes each node actively contends for channel access while at the same time is a potential receiver of some other node's transmission (i.e. the network

A. Helmy et al. (Eds.): MMNS 2006, LNCS 4267, pp. 61 – 73, 2006.

is saturated). A framework for conserving energy across routes in multi-hop and ad-hoc wireless networks is proposed in [3]. This approach adjusts the radio transmission power and the rebroadcast time of RREQ packets, using a rebroadcast mechanism for estimating the end-to-end energy consumption of a multi-hop network. These works have focused on energy consumption without any channel contention.

Recently, several researchers have begun to focus on the saturation throughput of DCF in error-prone channels [4-7]. To the best of our knowledge, most existing research only considers the error probability of transmission errors on data frames, and do not consider the influence of the channel state on the energy consumption. In [4], an improved analytical model is proposed that calculates IEEE 802.11a's DCF performance taking into account retransmitted packets and transmission errors. However, all the currently proposed performance models [4-6] ignore the influence of the physical layer and assume that the collision probability is same in all circumstances. The different types of frames transmitted by the different modulation schemes can result in different frame error rates. Furthermore, in a noisy wireless environment which is typically encountered in practice, all frames sent will have different error probabilities depending on the received signal strength. In [7], the performance model addresses the error probability of ACK frames. However, it only considers the error probability in the DCF basic access method.

On the other hand, some related work has been done on the analysis of energy consumption in DCF [2, 8, 9]. These models do not consider the transmission errors encountered in real wireless environments. In [10] and [11] the energy consumption models presented do consider the effect of transmission errors, however, the performance models address the effect of errors in data frames only (i.e. signaling and control frames are not considered). Furthermore, in [2, 10, 11], the impact of the EIFS interval has not been considered when a transmission failure occurs.

Therefore, in this paper, we first evaluate several critical PHY/MAC layer design components and their impact in an error-prone channel condition, and then we propose a novel model for energy consumption of an 802.11-based Ad-Hoc network in an error-prone wireless environment. In particular, we consider the impact of the noisy wireless channel and the number of competing nodes using both basic access and RTS/CTS exchange methods in IEEE 802.11-based DCF networks [12-15]. Results show that the transmitter produces much higher energy consumption when the channel is very noisy (e.g., when BER $\geq 10^{-5}$). The contribution of this research is to serve as an indication of the achievable reduction in energy consumption by using a cross-layer framework design for Ad-Hoc networks. A specific topology control, rate adaptation and routing strategy that takes advantage of this research will be reported in future research.

The rest of the paper is organized as follows. Section 2 gives an overview of the IEEE 802.11-based system. Section 3 presents the analytic model under different channel states for both basic access and RTS/CTS exchange methods in an error-prone wireless environment. Then we obtain the energy consumption model under the error-prone channel in section 4. Section 5 describes the simulation results and compares the energy consumption between 802.11a and g in DCF networks. Finally, we conclude the paper in Section 6.

2 System Overview

The IEEE 802.11 MAC [12] sub-layer provides a fairly controlled access to the shared wireless medium through two different access mechanisms: the basic access mechanism, called the distributed coordination function (DCF), and a centrally controlled access mechanism, called the point coordination function (PCF). In this paper, we focus on the energy consumption of IEEE 802.11 based networks using DCF.

2.1 IEEE 802.11a/b/g Physical Layer

The IEEE 802.11Physical (PHY) layer is an interface between the Medium Access Control (MAC) layer and the wireless medium, which transmits and receives data frames over the shared wireless medium. The exchange of frames between the MAC and PHY is under the control of the physical layer convergence procedure (PLCP) sub-layer. The PLCP protocol data units (PPDU) format of the IEEE 802.11 PHY includes PLCP preamble, PLCP header, PHY Sub-layer Service Data Units (PSDU), tail bits, and pad bits.

In the IEEE 802.11a/b/g PHY characteristics, we denote the PLCP preamble field, with duration of $tPLCPPreamble$. The PLCP header, with duration of $tPLCPHeader$. In IEEE 802.11a and g, the OFDM symbol interval denoted by T_m is 4μs. For IEEE 802.11b, we assume that $T_m = 1$. A summary of each frame type is given by equations (1)-(4) for both exchange methods (i.e. basic access and RTS/CTS). Note that, the FCS (frame check sequence) field in each MAC frame is a 32 bit cyclic redundancy code (CRC).

$$T_{r,m,data}(l) = tPLCPPreamble + tPLCPHeader + \frac{(MPDUHeader + FCS + l) \cdot 8}{D_{bps}(r)} \cdot T_m \quad (1)$$

$$T_{r,m,ack} = tPLCPPreamble + tPLCPHeader + \frac{ACKHeader/FCS \cdot 8}{S_{bps}(r)} \cdot T_m \quad (2)$$

$$T_{r,m,CTS} = tPLCPPreamble + tPLCPHeader + \frac{RTSHeader/FCS \cdot 8}{S_{bps}(r)} \cdot T_m \quad (3)$$

$$T_{r,m,RTS} = tPLCPPreamble + tPLCPHeader + \frac{CTSHeader/FCS \cdot 8}{S_{bps}(r)} \cdot T_m \quad (4)$$

$T_{r,m,data}(l)$ is the data transmission duration when a node transmits a data frame with l payload octets over a PHY scheme m using transmission rate r. $T_{r,m,ack}$, $T_{r,m,CTS}$ and $T_{r,m,RTS}$ denote the transmission times of Acknowledgement (ACK), Clear-to-send (CTS) and Ready-to-send (RTS) frames respectively, again for the selected transmission rate r and PHY scheme m. In IEEE 802.11 a/g, the basic rate set (BSS) is 6Mbps, 12 Mbps and 24 Mbps. Each station should support these rates and control information should be sent at these rates. We assume that all rates for IEEE 802.11 a/b/g can be used and $D_{bps}(r)$ and $S_{bps}(r)$ are the transmission rates of the data and control frames respectively.

2.2 The CSMA/CA Mechanism in DCF

The IEEE 802.11 MAC protocol's DCF, is based on carrier sense multiple access with collision avoidance (CSMA/CA). In operation a random backoff interval is uniformly chosen in the range $(0, W)$. The value W is called the contention window, which is an integer with range $CW_{m,min}$ (i.e. minimum contention window for PHY scheme m) to $CW_{m,max}$ (i.e. maximum contention window for PHY scheme m). A backoff counter is decremented while the medium is sensed idle, frozen when a transmission is detected on the channel, and reactivated when the channel is sensed idle again for more than a DIFS. The wireless node transmits when the backoff counter reaches zero. After each unsuccessful transmission, the contention window is doubled; up to a maximum backoff size $CW_{m,max}$. Since the backoff is uniformly distributed over $0,1,\ldots,CW_{m,min}$-1 for the first attempt, the backoff window size is $(CW_{m,mim}$-1)/2 on average. For a RTS/CTS exchange method, the short retry counter (SRC) is large than the maximum retry counter. The default values for LRC and SRC are 4 and 7 respectively. k' denotes the maximum retry counter, thus,

$$2^{k'} \cdot (CW_{m,min} + 1) = (CW_{m,max} + 1) \tag{5}$$

$W_m(k)$ is the average contention window for any selected PHY scheme m in a node A. After k consecutive unsuccessful transmission attempts this is given by

$$W_m(k) = \begin{cases} \dfrac{2^k \cdot (CW_{m,min} + 1) - 1}{2} & 0 \le k \le k' \\ \dfrac{CW_{m,max}}{2} & .k > k' \end{cases} \tag{6}$$

$T_{m,backoff}(k)$ is the average backoff interval in μsec for the PHY scheme m.

$$T_{m,backoff}(k) = W_m(k) * \sigma_m \tag{7}$$

k is the number of retransmission attempts. σ_m is the time duration of a slot in PHY scheme m.

3 Analysis of IEEE 802.11 a/b/g in an Error-Prone Channel

For an error-prone channel, unsuccessful transmission occurs not only when more than one node simultaneously transmits packets (contention), but also when poor channel conditions corrupt the packet. Since DCF has basic access and RTSCTS exchange methods, we assume packet errors occur from both data and ACK packets for basic access method and RTS, CTS, data and ACK packets for the RTS/CTS exchange method. For accurate analysis, we assume each node incurs a different packet error probability for its received packets. The packet error probability depends on the frame error rate (FER) when the packets are received.

The collision probability $P_{r,m,coll}$ is the probability that in a time slot at least one of the n-1 remaining nodes transmits. This is given by:

$$P_{r,m,coll} = 1 - (1 - \tau_m)^{n-1} \tag{8}$$

Where τ_m is the probability that a node transmits in a generic slot time.

$$
\tau_m = \begin{cases}
\dfrac{(1 - P_{r,m,coll}^{k+1}) \cdot 2 \cdot (1 - 2 \cdot P_{r,m,coll})}{W_m \cdot (1 - (2 \cdot P_{r,m,coll})^{k+1}) \cdot (1 - P_{r,m,coll}) + (1 - 2 \cdot P_{r,m,coll})(1 - P_{r,m,coll}^{k+1})} & k \le k' \\[4mm]
\dfrac{(1 - P_{r,m,coll}^{k+1}) \cdot 2 \cdot (1 - 2 \cdot P_{r,m,coll})}{\left(\begin{array}{l} W_m \cdot (1 - (2 \cdot P_{r,m,coll})^{k'+1}) \cdot (1 - P_{r,m,coll}) + (1 - 2 \cdot P_{r,m,coll}) \cdot (1 - P_{r,m,coll}^{k+1}) + \\ W_m \cdot 2^{k'} \cdot P_{r,m,coll}^{k'+1} \cdot (1 - 2 \cdot P_{r,m,coll}) \cdot (1 - P_{r,m,coll}^{k-k'}) \end{array} \right)} & k > k'
\end{cases}
\tag{9}
$$

Where $CW_{m,min}$ is the minimum backoff window size in m PHY scheme. After each unsuccessful transmission, the backoff window is doubled, up to a maximum backoff size. From equations (8) and (9), we define $P_{r,m,tr}$ as the probability in a slot time, of at least one or more transmissions. n active nodes contend to access the medium and each node has transmission probability τ_m.

$$
P_{r,m,tr} = 1 - (1 - \tau_m)^n
\tag{10}
$$

The probability of successful transmission $P_{r,m,s}$ is given by:

$$
P_{r,m,s} = \frac{n \cdot \tau_m \cdot (1 - \tau_m)^{n-1}}{P_{r,m,tr}} = \frac{n \cdot \tau_m \cdot (1 - \tau_m)^{n-1}}{1 - (1 - \tau_m)^n}
\tag{11}
$$

In basic access method, The transmission error probability ($P_{r,m,error}$) stands for the frame error rate (FER) of a MAC data frame or an ACK frame for a given node. We assume that the two events "data frame corrupted" and "ACK frame corrupted" are independent, and obtain:

$$
p_{r,m,error} = 1 - (1 - p_{r,m,data\_error}) \cdot (1 - p_{r,m,ack\_error})
\tag{12}
$$

Where $P_{r,m,data\_error}$ and P_{r,m,ack_error} are FERs of data frames and ACK frames for the selected transmission rate r and PHY scheme m respectively.

The $P_{r,m,error}$ for RTS/CTS exchange method is:

$$
p_{r,m,error} = 1 - (1 - p_{r,m,RTS\_error}) \cdot (1 - p_{r,m,CTS\_error}) \cdot \\
(1 - p_{r,m,data\_error}) \cdot (1 - p_{r,m,ack\_error})
\tag{13}
$$

Where P_{r,m,RTS_error}, P_{r,m,CTS_error}, $P_{r,m,data\_error}$ and P_{r,m,ack_error} are FERs of RTS frames, CTS frames, data frames and ACK frames for selected transmission rate r and PHY scheme m respectively.

The bit errors are uniformly distributed over the whole frame, the P_{r,m,RTS_error}, P_{r,m,CTS_error}, $P_{r,m,data\_error}$ and P_{r,m,ack_error} can then be calculated as:

$$
\begin{cases}
p_{r,m,data\_error} = 1 - (1 - BER_{m,data})^{N_{data}} \\
p_{r,m,ack\_error} = 1 - (1 - BER_{m,ack})^{N_{ack}}
\end{cases}
\begin{cases}
p_{r,m,RTS\_error} = 1 - (1 - BER_{m,RTS})^{N_{RTS}} \\
p_{r,m,CTS\_error} = 1 - (1 - BER_{m,CTS})^{N_{CTS}}
\end{cases}
\tag{14}
$$

Where N_{RTS}, N_{CTS}, N_{data} and N_{ack} are the length (bits) of the RTS, CTS, data and ACK frames and the $BER_{m,data}$ and $BER_{m,ack}$ are the bit error rates of the data and ACK frames respectively. The bit error rate is based on the selected PHY scheme, which is determined by SNR, modulation scheme and code scheme or transmission rate. Thus, we can obtain the received FER by measuring the SNR from a PLCP Protocol Data Unit (PPDU) packet in transmission rate r in PHY scheme m. The bit error rate (BER) can then be estimated by measuring the bit-energy-to-noise ratio [16].

$p'_{r,m,s}$ denotes the successful transmission probability for a single station in an error-prone channel condition, which is defined as the probability that only one of n nodes is successfully transmitting and there are no corrupted packets. The probability of one node successfully transmitting in error-prone channel for the basic access method is:

$$P'_{r,m,s} = P_{r,m,s} \cdot (1 - P_{r,m,error}) \tag{15}$$

From equation (12), thus we could denote the probability for one node successfully transmitting in error-prone channel as shown in following,

$$P'_{r,m,s} = P_{r,m,s} \cdot (1 - P_{r,m,data\_error}) \cdot (1 - P_{r,m,ack\_error}) \tag{16}$$

Let $P_{r,m,dataframe\_error}$ stands for the probability that a transmission error occurs on a data frame in a time slot; this occurs when one and only one station transmits in a time slot and the data frame is corrupted because of transmission errors. $P_{r,m,ackframe\_error}$ denotes the probability that a data frame transmission is successful but the corresponding ACK frame is corrupted due to transmission errors.

Thus, $P_{r,m,dataframe\_error}$ and $P_{r,m,ackframe\_error}$ can be expressed using the following equations:

$$P_{r,m,dataframe\_error} = P_{r,m,s} \cdot P_{r,m,data\_error} \tag{17}$$

$$P_{r,m,ackframe\_error} = P_{r,m,s} \cdot (1 - P_{r,m,data\_error}) \cdot P_{r,m,ack\_error} \tag{18}$$

Otherwise, from equation (13), the $P'_{r,m,s}$ for the RTS/CTS exchange method is:

$$P'_{r,m,s} = P_{r,m,s} \cdot (1 - p_{r,m,RTS\_error}) \cdot (1 - p_{r,m,CTS\_error}) \cdot (1 - p_{r,m,data\_error}) \cdot$$
$$(1 - P_{r,m,ack\_error})$$

$$P_{r,m,RTSframe\_error} = P_{r,m,s} \cdot P_{r,m,RTS\_error}$$

$$P_{r,m,CTSframe\_error} = P_{r,m,s} \cdot (1 - P_{r,m,RTS\_error}) \cdot P_{r,m,CTS\_error} \tag{19}$$

$$P_{r,m,dataframe\_error} = P_{r,m,s} \cdot (1 - P_{r,m,RTS\_error}) \cdot (1 - P_{r,m,CTS\_error}) \cdot P_{r,m,data\_error}$$

$$P_{r,m,ackframe\_error} = P_{r,m,s} \cdot (1 - P_{r,m,RTS\_error}) \cdot (1 - P_{r,m,CTS\_error}) \cdot$$
$$(1 - P_{r,m,data\_error}) \cdot P_{r,m,ack\_error}$$

$P_{r,m,RTSframe\_error}$, $P_{r,m,CTSframe\_error}$, $P_{r,m,dataframe\_error}$ and $P_{r,m,ackframe\_error}$ are the respective probability of a transmission error occurring in a RTS, CTS, data or ACK frame.

4   Energy Consumption of IEEE 802.11/ a/b/g in an Error-Prone Channel

In the IEEE 802.11 DCF, two power management mechanisms are supported: active and power saving mechanism (PSM). In this paper, we only consider the active mechanism, in which a node may be in one of three different radio modes, namely, transmit, receive, and idle modes. The main parameters of our energy model are:

- $P_{r,m,tx}$: the energy required for the transmission rate r and PHY scheme m to transmit data to the destination (mJ/sec).

- $P_{r,m,rx}$: the energy required for the transmission rate r and PHY scheme m to receive data from source (mJ/sec).
- $P_{r,m,sense}$: the energy required for a PHY scheme m to sense that the radio signal (mJ/sec) is idle.

The timing of successful two-way and four-way frame exchanges is shown in Table 1 and 2. Alternatively, if an RTS (ACK) frame is not received, for example because of an erroneous reception of the preceding CTS (Data) frame, the transmitter will contend for the medium to re-transmit the frame after a CTS (ACK) timeout.

Table 1. Channel states for the basic access method under DCF

No.	Scenario	T	Duration
1	Idle	$T_{r,m,idle}$	σ_m
2	Success	$T_{r,m,s}(l)$	$T_{r,m,data}(l)+T_{r,m,SIFS}+T_{r,m,ack}+T_{r,m,DIFS}+2*\delta_m$
3	Collision	$T_{r,m,c}(l)$	$T_{r,m,data}(l)+T_{r,m,EIFS}+\delta_m$
4	Data corruption	$T_{r,m,data\_error}(l)$	$T_{r,m,data}(l)+T_{r,m,EIFS}+\delta_m$
5	ACK corruption	$T_{r,m,ack\_error}(l)$	$T_{r,m,data}(l)+T_{r,m,ACK\_timeout}+T_{r,m,DIFS}+2*\delta_m$

Table 2. Channel states for the RTS/CTS exchange method under DCF

No.	Scenario	T	Duration
1	Idle	$T_{r,m,idle}$	σ_m
2	Success	$T_{r,m,s}(l)$	$T_{r,m,RTS}+T_{r,m,CTS}+T_{r,m,data}(l)+T_{r,m,ack}$ $+T_{r,m,DIFS}+3*T_{r,m,SIFS}+4*\delta_m$
3	Collision	$T_{r,m,c}(l)$	$T_{r,m,RTS}+T_{r,m,EIFS}+\delta_m$
4	RTS corruption	$T_{r,m,RTS\_error}(l)$	$T_{r,m,RTS}+T_{r,m,EIFS}+\delta_m$
5	CTS corruption	$T_{r,m,CTS\_error}(l)$	$T_{r,m,RTS}+T_{r,m,CTS\_timeout}+T_{r,m,DIFS}+2*\delta_m$
6	Data corruption	$T_{r,m,data\_error}(l)$	$T_{r,m,RTS}+T_{r,m,CTS}+T_{r,m,data}(l)+T_{r,m,EIFS}$ $+2*T_{r,m,SIFS}(m)+3*\delta_m$
7	ACK corruption	$T_{r,m,ack\_error}(l)$	$T_{r,m,RTS}+T_{r,m,CTS}+T_{r,m,data}(l)+T_{r,m,ack\_timeout}$ $+T_{r,m,DIFS}+2*T_{r,m,SIFS}(m)+4*\delta_m$

$T_{r,m,EIFS}$ is derived from the SIFS and the DIFS and the length of time it takes to transmit an ACK (CTS) Control frame at control signal transmission rate by the following equation:

$$T_{m,EIFS} = T_{r,m,SIFS} + T_{r,m,ack} + T_{r,m,DIFS} \qquad (20)$$

$$T_{r,m,ack\_timeout} = T_{r,m,SIFS} + T_{r,m,ack} \qquad (21)$$

$$T_{r,m,CTS\_timeout} = T_{r,m,SIFS} + T_{r,m,CTS} \qquad (22)$$

We assume that, under a heavy traffic load (i.e. saturation), n competing nodes always have packets to transmit. On average, node A incurs collisions $N_{m,collision}$ times before a successful transmission. $P_{r,m,collision}$ is the collision probability. Thus, $(1-P_{r,m,collision})$ is the probability of probability of successful transmission. Therefore, the average number of retransmissions before a successful transmission is $1/(1-P_{r,m,collision})$. The number of collisions is

$$N_{m,collision} = \frac{1}{1-(P_{r,m,collision} + (1-P_{r,m,collision}) \cdot P_{r,m,error})} - 1 \tag{23}$$

Let $E_{m,collision}(l)$ is the energy consumption in collision period in node A.
In the basic access method, the $E_{m,collision}(l)$ is,

$$E_{r,m,collision}(l) = N_{m,collision} \cdot (T_{r,m,data}(l) \cdot P_{r,m,tx} + (T_{r,m,EIFS} + \delta_m) \cdot P_{r,m,sense}) \tag{24}$$

The $E_{m,collision}(l)$ for the RTS/CTS exchange method is

$$E_{r,m,collision}(l) = N_{m,collision} \cdot (T_{r,m,RTS} \cdot P_{r,m,tx} + (T_{r,m,EIFS} + \delta_m) \cdot P_{r,m,sense}) \tag{25}$$

During the backoff period of node A, it will overhear transmissions from other nodes. The average number overheard in A is denoted as $N_{m,overheard}$.

$$N_{m,overheard} = W_m(k) \cdot P_{r,m,tr} \tag{26}$$

Among the $N_{m,overheard}$ transmissions, the probability of a successful transmission is $P_{r,m,s}$, and $(1-P_{r,m,s})$ is the probability of unsuccessful transmission. $E_{r,m,overheard}(l)$ denotes the energy consumption for transmissions overheard in node A.
In the basic access method this is,

$$E_{r,m,overheard}(l) = N_{m,overheard} \cdot P_{r,m,rx} \cdot$$
$$\begin{pmatrix} P'_{r,m,s} \cdot P_{r,m,tr} \cdot T_{r,m,s}(l) + P_{r,m,tr} \cdot (1-p'_{r,m,s}) \cdot T_{r,m,c}(l) + \\ P_{r,m,dataframe\_error} \cdot T_{r,m,data\_error}(l) + P_{r,m,ackframe\_error} \cdot T_{r,m,ack\_error}(l) \end{pmatrix} \tag{27}$$

For the RTS/CTS exchange method this is,

$$E_{r,m,overheard}(l) = N_{m,overheard} \cdot P_{r,m,rx} \cdot$$
$$\begin{pmatrix} P'_{r,m,s} \cdot P_{r,m,tr} \cdot T_{r,m,s}(l) + P_{r,m,tr} \cdot (1-p'_{r,m,s}) \cdot T_{r,m,c}(l) + \\ P_{r,m,RTSframe\_error} \cdot T_{r,m,RTS\_error}(l) + P_{r,m,CTSframe\_error} \cdot T_{r,m,CTS\_error}(l) + \\ P_{r,m,dataframe\_error} \cdot T_{r,m,data\_error}(l) + P_{r,m,ackframe\_error} \cdot T_{r,m,ack\_error}(l) \end{pmatrix} \tag{28}$$

We assume that a node will remain in the sensing mode if it doesn't transmit, node A will therefore remain in the sensing mode during all the backoff periods, and it will backoff for $N_{m,backoff}$ times. Thus,

$$N_{m,backoff} = \frac{1}{1-(P_{r,m,collision} + (1-P_{r,m,collision}) \cdot P_{r,m,error})} \tag{29}$$

Let $E_{r,m,backoff}$ is the energy consumption in the backoff period in node A.

$$E_{r,m,backoff} = N_{m,backoff} \cdot T_{m,backoff}(k) \cdot P_{r,m,sense} \tag{30}$$

We assume that $E_{r,m}(l)$ is the overall energy consumption of a successful data frame transmission, with length packet l ,over the selected transmission r rate and PHY scheme m.

$$E_{r,m}(l) = E_{r,m,backoff} + E_{r,m,overheard}(l) + E_{r,m,collision}(l) + E_{r,m,error}(l) + E_{r,m,transmission}(l) \tag{31}$$

$E_{r,m,backoff}$ is the energy consumption during a backoff period in a transmitter. $E_{r,m,collision}(l)$ is the energy consumption in a transmitter during a packet collision. The average number overheard in A is denoted as $N_{m,overheard}$. $E_{r,m,transmission}(l)$ is the total energy consumption during a successful transmission when there is no collision or packet errors. Thus, the $E_{r,m,transmission}(l)$ and $E_{r,m,error}(l)$ in Basic access method are,

$$E_{r,m,transmission} = T_{r,m,data}(l) \cdot P_{r,m,tx} + T_{r,m,ack} \cdot P_{r,m,rx} + (T_{r,m,SIFS} + T_{r,m,DIFS} + 2 \cdot \delta_m) \cdot P_{r,m,sense}$$

$$E_{r,m,error} = N_{r,m,dataframe\_error} \cdot \left(\begin{array}{c} T_{r,m,data}(l) \cdot P_{r,m,tx} + \\ (T_{r,m,EIFS} + \delta_m) \cdot P_{r,m,sense} \end{array} \right) + \tag{32}$$

$$N_{r,m,ackframe\_error} \cdot \left(\begin{array}{c} T_{r,m,data}(l) \cdot P_{r,m,tx} + \\ (T_{r,m,ACK\_timeout} + T_{r,m,DIFS} + 2 \cdot \delta_m) \cdot P_{r,m,sense} \end{array} \right)$$

For the RTS/CTS exchange method this is,

$$E_{r,m,transmission} = (T_{r,m,RTS} + T_{r,m,data}(l)) \cdot P_{r,m,tx} + (T_{r,m,CTS} + T_{r,m,ack}) \cdot P_{r,m,rx} + (T_{r,m,DIFS} + 3 \cdot T_{r,m,SIFS} + 4 \cdot \delta_m) \cdot P_{r,m,sense}$$

$$E_{r,m,error} = N_{r,m,RTSframe\_error} \cdot \left(T_{r,m,RTS} \cdot P_{r,m,tx} + (T_{r,m,EIFS} + \delta_m) \cdot P_{r,m,sense} \right) +$$

$$N_{r,m,CTSframe\_error} \cdot \left(\begin{array}{c} T_{r,m,RTS} \cdot P_{r,m,tx} + (T_{r,m,CTS\_timeout} + \\ T_{r,m,DIFS} + 2 \cdot \delta_m) \cdot P_{r,m,sense} \end{array} \right) +$$

$$N_{r,m,dataframe\_error} \cdot \left(\begin{array}{c} (T_{r,m,RTS} + T_{r,m,data}(l)) \cdot P_{r,m,tx} + T_{r,m,CTS} \cdot P_{r,m,rx} + \\ (T_{r,m,EIFS} + 2 \cdot T_{r,m,SIFS} + 3 \cdot \delta_m) \cdot P_{r,m,sense} \end{array} \right) + \tag{33}$$

$$N_{r,m,ackframe\_error} \cdot \left(\begin{array}{c} (T_{r,m,RTS} + T_{r,m,data}(l)) \cdot P_{r,m,tx} + T_{r,m,CTS} \cdot P_{r,m,rx} + \\ (T_{r,m,ack\_timeout} + T_{r,m,DIFS} + 2 \cdot T_{r,m,SIFS} + 4 \cdot \delta_m) \cdot \\ P_{r,m,sense} \end{array} \right)$$

$N_{r,m,RTSframe\_error}$, $N_{r,m,CTSframe\_error}$, $N_{r,m,dataframe\_error}$ and $N_{r,m,ackframe\_error}$ are the number of transmissions error occurring in RTS, CTS, data and ACK frames respectively.

$$\begin{cases} N_{r,m,dataframe\_error} = \dfrac{1}{1 - P_{r,m,dataframe\_error}} \\ N_{r,m,ackframe\_error} = \dfrac{1}{1 - P_{r,m,ackframe\_error}} \end{cases} \begin{cases} N_{r,m,RTSframe\_error} = \dfrac{1}{1 - P_{r,m,RTSframe\_error}} \\ N_{r,m,CTSframe\_error} = \dfrac{1}{1 - P_{r,m,CTSframe\_error}} \end{cases} \tag{34}$$

Therefore, the energy consumption per bit $G_{r,m}(l)$ can be computed as:

$$G_{r,m}(l) = \frac{E_{r,m}(l)}{l \cdot 8} \tag{35}$$

5 Simulation Results and Discussion

A simulation environment was developed using the Qualnet simulation Lab [17]. Although extensive simulations have been conducted, in this paper, we only present a summary of those results that focus on energy consumption in IEEE 802.11-based networks. In the model, a range of randomly distributed errors were introduced into the channel. The specific BER values were chosen as they are the values used by commercial wireless adapters for rate adaptation (e.g. Orinoco) [18].

Table 3. The energy consumption and MAC parameters used in simulation. [19]

	Transmit	Receive	Standby	$CW_{m,min}$	$CW_{m,max}$	$aSlotTime$ (σ_{m})
802.11a	554 mA	318 mA	203 mA	16 slots	1024 slots	9 µs
802.11b	539 mA	327 mA	203 mA	32 slots	1024 slots	20 µs
802.11g	530 mA	282 mA	203 mA	32 slots	1024 slots	20 µs

To observe the energy consumption in a relatively heavy loaded network, we set up a simulation scenario with 30 active nodes during the channel competition. The transmission bit rate is 6 Mbps in both the basic access and RTS/CTS exchange methods for the 802.11a and g. Each active node transmits 2304 bytes of traffic in their data frames.

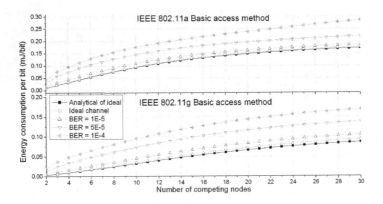

Fig. 1. Energy consumption per bit of basic access method for 802.11a/g

Figures 1 and 2 show the energy consumption per bit (mJ/bit) when transmitting using the basic access and RTS/CTS exchange methods. The results consider 802.11a/OFDM and 802.11g/ERP-OFDM PHYs with different bit error rates. Analytical results for an ideal channel are also shown to agree well with the simulations. The results show that energy consumption is primarily influenced by the selected MAC and PHY layers. The 802.11a with basic access method is most sensitive to channel interference; and 802.11g consumes less energy than 802.11a at the same data rate for both methods. The RTS/CTS exchange method has lower energy consumption than the basic access method in both 802.11a and g. This is

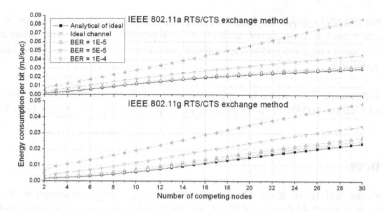

Fig. 2. Energy consumption per bit of RTS/CTS exchange method for 802.11a/g

because, when the number of competing nodes is increased, the congestion is also increased; the cost of packet retransmissions is also lower in the RTS/CTS exchange method than in the basic access method. Furthermore, for an ideal channel, the results also show that a MAC sub-layer with a larger minimum contention window size (e.g. 802.11g has a lager $CW_{m,min}$ of 31 slots than 802.11a of 15 slots in MAC sub-layer protocol) will reduce the energy consumption under saturation conditions for both exchange methods.

Another interesting result in this study is the energy consumption under the varying bit error rates. The results show that the transmitter produces much higher energy consumption when the channel is very noisy (e.g., when BER$\geq$10^{-5}). The bulk of frames are dropped due to transmission errors. This is because the larger frame size has a higher probability of transmission error when the channel condition is noisy. Therefore there will be more data frames corrupted by the channel thus the higher cost of energy consumption per bit. There will be less RTS/CTS packets corrupted as they are smaller sized than the data frames; hence the results show that although the RTS/CTS method consumes less energy than the basic access method, it does appear to be more sensitive when used in an 802.11a network operating in noisy conditions.

6 Conclusions

This paper presents a new approach for predicting energy consumption in IEEE 802.11-based DCF networks with multiple modulation and PHY layers. The energy consumption model developed may be applied to both congested and error-prone wireless channels. Analysis shows that the selected MAC and PHY layers have the primary influence on the energy consumption. The simulation results are observed to match very well with the analytical results, which show that the condition of the wireless channel and the level of congestion both increase the energy consumption. Moreover, this study shows that the energy consumption of the RTS/CTS exchange method is less than that of the basic access method. This paper describes research that aims to model energy consumption in IEEE-based DCF networks under an error-prone channel condition. An interesting area of future research will be to extend the approach into a cross-layer architecture to provide further rate adaptation to optimize network capacity and reduce energy consumption in 802.11-based Multi-hop wireless networks.

Acknowledgment. This work is supported by the UK funding body EPSRC, under the project GR/S02105/01 "Programmable Routing Strategies for Multi-Hop Wireless Networks".

References

1. M. Stemm and R. H. Katz: Measuring and reducing energy consumption of network interfaces in handheld devices: ICICE Trans. On Communications, 8, pp 1125-1131, 1997.
2. Carvalho, M.M.; Margi, C.B.; Obraczka, K.; Garcia-Luna-Aceves, J.J.: Modeling energy consumption in single-hop IEEE 802.11 ad hoc networks: Computer Communications and Networks, 2004. ICCCN 2004. Proceedings. 13th International Conference, pp 367- 372, 11-13 Oct. 2004.
3. T.H. Lee, A. Marshall and B. Zhou: A Framework for Cross-layer Design of Energy-conserving On-Demand Routing in Multi-hop Wireless Networks: IEE Mobility Conference 2005, The Second International Conference on Mobile Technology, Applications and Systems, Nov 15-17, 2005, Guangzhou, China.
4. P. Chatzimisios, A. C. Boucouvalas, and V. Vitsas: Performance Analysis of IEEE 802.11 DCF in the Presence of Transmission Errors: IEEE ICC2004, Paris, June 2004.
5. V.M. Vishnevsky and A.I.Lyakhov: 802.11 LANs: Saturation Throughput in the Presence of Noise: Proc. of 2nd Int. IFIP TC6 Networking Conf., Pisa, Italy, May, 2002.
6. Jihwang Yeo; Agrawala, A.: Packet error model for the IEEE 802.11 MAC protocol: Personal, Indoor and Mobile Radio Communications, 2003. PIMRC 2003. 14th IEEE Proceedings on , vol.2, no.pp. 1722- 1726 vol.2, 7-10 Sept. 2003.
7. Qiang Ni, Tianji Li, Thierry Turletti, and Yang Xiao: Saturation Throughput Analysis of Error-Prone 802.11 Wireless Networks: Wiley Journal of Wireless Communications and Mobile Computing (JWCMC), Vol. 5, Issue 8, Dec. 2005, pp. 945-956.
8. L. Bononi, M. Conti, and L. Donatiello: A Distributed Mechanism for Power Saving in IEEE 802.11 Wireless LANs: Mobile Networks and Applications 6,211-222, 2001.
9. Cunha, D. O., Costa, L. H. M. K., and Duarte, O.: Analyzing the Energy Consumption of IEEE 802.11 Ad Hoc Networks: 6th IFIP IEEE International Conference on Mobile and Wireless Communication Networks, MWCN'2004, Paris, France, October 2004.
10. Xiaodong Wang, Jun Yin, and Dharma P. Agrawal: Analysis and Optimization of the Energy Efficiency in 802.11 DCF: ACM/Kluwer Journal on Mobile Networks and Applications (MONET) Special Issue on Internet Wireless Access: 802.11 and Beyond, vol. 11, no. 3, June 2005.
11. Daji Qiao, Sunghyun Choi, Amit Jain, and Kang G. Shin: MiSer: An Optimal Low-Energy Transmission Strategy for IEEE 802.11 a/h: Proc. ACM MobiCom'2003, San Diego, CA, September, 2003.
12. IEEE 802.11 Work Group: Part 11: Wireless LAN Medium Access Control (MAC) and Physical Layer (PHY) Specifications: ANSI/IEEE Std 802.11, 1999.
13. IEEE 802.11 Work Group: Part 11: Wireless LAN Medium Access Control (MAC) and Physical Layer (PHY) specification: High-speed Physical Layer Extension in the 5 GHz Band: ANSI/IEEE Std 802.11a, 1999.

14. IEEE 802.11 Work Group: Part 11: Wireless LAN Medium Access Control (MAC) and Physical Layer (PHY) specification: High-speed Physical Layer Extension in the 2.4 GHz Band: ANSI/IEEE Std 802.11b, 1999.
15. IEEE 802.11 Work Group: Part 11: Wireless LAN Medium Access Control (MAC) and Physical Layer (PHY) specifications: Further Higher Data Rate Extension in the 2.4 GHz Band: ANSI/IEEE Std 802.11g, 1999.
16. Rappaport TS.: Wireless Communications: Principles and Practice: Prentice Hall, New Jersey, USA, 1996.
17. Scalable Networks, http://www.scalable-networks,com
18. Proxim Orinoco Wireless LAN, http://www.proxim.com
19. Cisco Aironet 802.11a/b/g Wireless LAN Client, http://www.cisco.com

MaVIS: Media-aware Video Streaming Mechanism

Sunhun Lee and Kwangsue Chung

School of Electronics Engineering, Kwangwoon University, Korea
sunlee@adams.kw.ac.kr, kchung@kw.ac.kr

Abstract. Existing streaming protocols have no consideration for the characteristics of streaming applications because they only consider the network stability. In this paper, in order to overcome limitations of the previous schemes for video streaming, we propose a new video streaming mechanism called "MaVIS (Media-aware Video Streaming)". The MaVIS takes more sophisticated mechanism that considers both network and user requirements. Proposed mechanism improves the network stability by adjusting the sending rate suitable for current network state and it also provides the smoothed playback by preventing buffer underflow or overflow. Moreover, it is designed to consider characteristics of the video stream. Through the simulation, we prove that the MaVIS mechanism efficiently uses the buffer resources and provides improved network stability and smoothed playback.

1 Introduction

The Internet has recently been experiencing an explosive growth in the use of audio and video streaming applications. Such applications are delay-sensitive, semi-reliable and rate-based. Thus they require isochronous processing and QoS (Quality-of-Service) from the end-to-end viewpoint. However, today's Internet does not attempt to guarantee an upper bound on end-to-end delay and a lower bound on available bandwidth. As a result, most of video streaming applications use UDP (User Datagram Protocol) as a transport protocol that has no congestion control mechanism. However, the emergence of non-congestion-controlled streaming applications threatens unfairness to competing TCP (Transmission Control Protocol) traffic and possible congestion collapse [1].

Studies on the congestion controlled streaming protocol has been increasingly done since the 1990s [2, 3, 4, 5, 6, 7, 8, 9, 10, 11, 12, 13]. These works attempt to guarantee the network stability and fairness with competing TCP traffic. However, by considering only the network stability, these works ignore the quality of video stream which is served to the end user. Moreover, most of existing streaming protocols have no consideration for the characteristics of video stream which affects significantly on the quality of streaming services.

In this paper, we propose a new video streaming mechanism called "MaVIS (Media-aware Video Streaming)". The MaVIS adjusts sending rate of video stream based on the current network state and controls the quality of video

A. Helmy et al. (Eds.): MMNS 2006, LNCS 4267, pp. 74–84, 2006.

stream based on the receiver's buffer occupancy. It also considers the media characteristics of video stream which is requested from the end user. Therefore, the MaVIS maintains the network stability accomplished by previous works and achieves the smoothed playback by preventing the buffer underflow or overflow.

The rest of this paper is organized as follows. In Section 2, we review and discuss some of the previous works and in Section 3, we present the concept and algorithms introduced in the MaVIS mechanism. Detailed description of our simulation results are presented in Section 4. Finally, Section 5 concludes the paper and discusses some of our future works.

2   Related Works

While data applications such as Web and FTP (File Transfer Protocol) are mostly based on TCP, multimedia streaming applications will be based on UDP due to its real-time characteristics. However, UDP does not support congestion control mechanism. For this reason, UDP-based streaming protocols cause the unfairness with competing TCP traffic and the starvation of congestion controlled TCP traffic which reduces its bandwidth share during overload situation. To overcome this limitation, several congestion controlled streaming protocols have been developed, recently. These works mostly interested in network stability. For the network stability, they apply the congestion control mechanism to the non-congestion controlled streaming protocol, such as UDP, or modify the existing congestion control mechanism in consideration of real-time characteristics of video streaming.

Previous works, interested in the network stability, can be further categorized into three different approaches. The first approach mimics TCP's AIMD (Additive Increase Multiplicative Decrease) algorithm to achieve the TCP-friendliness [2, 3, 4, 5, 6]. The second approach adjusts its sending rate according to the Padhye's TCP throughput modeling equation [7, 8, 9, 10]. The third approach uses the bandwidth estimation scheme that was employed in the TCP Westwood [11, 12].

A typical example is the TFRC (TCP-Friendly Rate Control) for unicast streaming applications which uses the TCP throughput modeling equation [14]. Padhye et al. presents an analytical model for the available bandwidth share(T) of TCP connection with S as the segment size, p as the packet loss rate, t_{RTT} as the RTT (Round Trip Time), t_{RTO} as the RTO (Retransmission Time-Out). The average bandwidth share of TCP connection depends mainly on t_{RTO} and p as shown in (1):

$$T = \frac{S}{t_{RTT}\sqrt{\frac{2p}{3}} + t_{RTO}(3\sqrt{\frac{3p}{8}})p(1 + 32p^2)} \qquad (1)$$

This approximation is reasonable for reaching TCP-friendliness. For this reason, many TCP-friendly rate control schemes use this analytical model.

The TFRC with Padhye's modeling equation is responsive to network congestion over longer time period and changes the sending rate in a slowly responsive manner. Therefore, it can work well with TCP flows and reduce the packet loss occurrence. But it has some limitations which are oscillations of sending rate and possibility of buffer underflow or overflow. It also has no consideration for the quality of streaming service. The improved network stability does not guarantee an improved quality of service on video streaming.

Unlike the network viewpoint protocol, discussed above, the buffer-driven scheme is a streaming protocol which provides smoothed playback of video stream to the user. Usually, streaming applications use a buffer to compensate delay and jitter caused from the network. The buffer-driven scheme controls the video quality and schedules data transmission based on receiver and sender buffer occupancies. Therefore, it provides smoothed playback to the user by preventing buffer underflow or overflow. However, the buffer-driven scheme still has instability and unfairness problems because it has no concern about the network state [13].

Figure 1 shows the existing streaming protocols which have been classified into four different approaches. Existing streaming protocols control the transmission rate of the video stream by using (1) the modified congestion control mechanism, (2) the Padhye's TCP throughput modeling equation, (3) the estimation of available network bandwidth, and (4) the buffer occupancy. The MaVIS mechanism, proposed in this paper, takes two approaches together to achieve both the network stability and the smoothed playback, simultaneously.

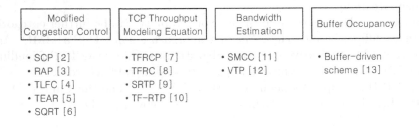

Fig. 1. Classification of existing streaming protocols

3 MaVIS Mechanism

3.1 Overall Architecture

The existing streaming protocols mostly interested in network stability. To improve the network stability, they control the sending rate of video stream according to the current network state. However, for the efficient video streaming service over Internet, streaming protocol has to provide the smoothed playback to the user by preventing buffer underflow or overflow. Moreover, to improve the quality of streaming service, it has to consider the media characteristics of video stream.

We propose a new video streaming mechanism that controls the sending rate of video stream based on current network state and adjusts the video quality based on current receiver buffer occupancy. It also efficiently manages the buffer resource on the basis of the characteristics of video stream, typically an encoding rate of video stream. Figure 2 shows the overall architecture of the proposed mechanism. Basically, the MaVIS mechanism relies on the RTP/RTCP (Realtime Transport Protocol/Realtime Transport Control Protocol) model which is standardized by IETF (Internet Engineering Task Force). The RTCP periodically reports the information on network state to the sender [15]. According to this feedback information, the MaVIS determines the network state as congestion_state or stable_state.

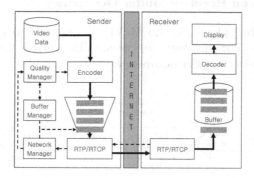

Fig. 2. Overall architecture

After deciding the network state, the MaVIS estimates the receiver buffer occupancy at a sender side. Then it controls the sending rate or the video quality on the basis of the current network state and buffer occupancy. If the estimated receiver buffer occupancy is very high or low, then it controls the video quality to prevent buffer underflow or overflow. Otherwise, it controls the sending rate for network stability according to the current network state.

3.2   Decision on Network State

The MaVIS decides the network state based on the packet loss rate which can be calculated from the RR (Receiver Report) of RTCP. The packet loss rate is estimated in a receiver side by checking the sequence numbers included in RTP header of the data packets. It is calculated on the proportion of the number of packets received to the number of packets sent during the RTCP interval [15].

$$\text{Packet loss rate}, p = 1 - \frac{\text{Number of packets received}}{\text{Number of packets sent}} \tag{2}$$

Figure 3 shows the decision rule for network state. If the packet loss rate is zero, the network state is decided to be stable_state. Otherwise, the network state is decided to be congestion_state by assuming that the packet losses are caused by congestion.

If (Packet loss rate = 0)
  Network state = stable_state
Else (Packet loss rate > 0)
  Network state = congestion_state

Fig. 3. Decision rule for network state

3.3 Estimation on Receiver Buffer Occupancy

To provide smoothed playback, the MaVIS is designed to prevent buffer underflow or overflow. For this purpose, the MaVIS controls the video quality by estimating the receiver's buffer occupancy.

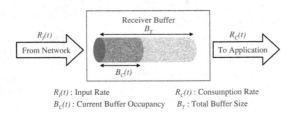

$R_I(t)$: Input Rate $R_C(t)$: Consumption Rate
$B_C(t)$: Current Buffer Occupancy B_T : Total Buffer Size

Fig. 4. Buffer state in a receiver side

Figure 4 shows the buffer state in a receiver side with B_T as the total buffer size, $B_C(t)$ as the current buffer occupancy, $R_I(t)$ as the input data rate in a receiver. If we ignore the network delay, $R_I(t)$ is equal to the outgoing rate in a sender. $R_C(t)$ is the consumption rate by streaming application at a receiver side, equal to the encoding rate of video stream. The sender can estimate the receiver buffer occupancy(B_E) as follows:

$$B_E(t+1) = B_C(t) + \{R_I(t) - R_C(t)\} \times RTT \qquad (3)$$

$$B_C(t) \approx B_E(t) \qquad (4)$$

$$R_I(t) = \frac{\text{Number of packets} \times \text{Packet size}}{\text{RTCP interval}} \qquad (5)$$

$$R_C(t) = \text{Encoding rate at time (t)} \qquad (6)$$

If the receiver buffer occupancy is estimated to be underflow or overflow, encoding rate of video stream is changed. After changing the next video quality, $B_C(t)$ is re-calculated based on the changed encoding rate. In (3), the MaVIS

Table 1. Control scheme in MaVIS

Buffer occupancy	Network state	
	congestion_state	stable_state
$0 \sim MinTH\%$	Decrease video quality	Decrease video quality
$MinTH \sim MaxTH\%$	Decrease rate by α	Increase rate by β
$MaxTH \sim 100\%$	Increase video quality	Increase video quality

predicts one-RTT-early the receiver buffer occupancy. This approach efficiently prevents buffer underflow or overflow with no additional overhead in high delay networks.

The management of buffer resource in MaVIS takes a characteristic of video stream into consideration. Encoding rate of video stream, determined by using a video compression scheme, is a representative characteristic of video stream, because it has an effect on change of network conditions and buffering level. To efficiently manage the buffer resource, MaVIS defines two important thresholds which prevent the buffer underflow or overflow. These thresholds divide the operation of MaVIS between the response to network state and the response to buffer occupancy. A threshold with no consideration of the encoding rate causes an unnecessary change of sending rate and video quality that can be annoying to the user. To overcome this problem, the MaVIS sets these thresholds based on the encoding rate of video stream. These appropriate thresholds will help to improve efficiency of buffer resources. It also reduces unnecessary change of sending rate and video quality. The MaVIS sets the thresholds as shown in (7). R_{ENC} is the encoding rate of video stream and B_T is total receiver buffer size. These two thresholds are set to the sufficient buffering level which does not cause either buffer underflow or overflow for one second.

$$MinTH(\%) = \frac{R_{ENC}(bps)/8bits}{B_T(bytes)} \times 100, MaxTH(\%) = 100 - MinTH \quad (7)$$

3.4   Sending Rate and Video Quality Control

The MaVIS firstly determines the network state as congestion_state or stable_state based on the packet loss rate. And then it estimates the receiver buffer occupancy. Based on the network state and the buffer occupancy, it either controls the sending rate or the video quality.

Table 1 shows the control scheme that adjusts sending rate or video quality based on network state and buffer occupancy. For the congestion_state, the sender has to reduce the sending rate for the network stability. But, if the receiver buffer occupancy is estimated as $(0 \sim MinTH\%)$, we decreased the video quality to prevent buffer underflow without adjusting the sending rate. For $(MaxTH \sim 100\%)$ of buffer occupancy, the video quality is increased to prevent the buffer overflow with no sending rate control. However, in case of $(MinTH \sim MaxTH\%)$, it simply reduces the sending rate by α for the network stability.

For $(0 \sim MinTH\%)$ of the buffer occupancy in the stable-state, the sender only decreases the video quality. However, for $(MaxTH \sim 100\%)$, it increases the video quality. In case of $(MinTH \sim MaxTH\%)$, the MaVIS scheme increases the sending rate by β for competing with other traffics.

Parameter α and β are derived from the congestion control mechanism need in conventional TCPs, to improve fairness with competing TCP traffics. From [8], it is found that the overall TCP bandwidth share is inversely proportional to the square root of the packet loss rate. Hence, after receiving a loss notification from the receiver, the sender can reduce its sending rate by α from the current sending rate, R_C.

$$\alpha = R_C \times \sqrt{p} \tag{8}$$

For the case of stable-state, the sending rate can be increased by parameter, β. This increase value does not exceed the increase of the competing TCP connection under the same network conditions. Thus, the sender has to adjust the sending rate to be equal to the competing TCP connection, in order to improve the fairness. For this requirement, the MaVIS scheme increases sending rate by R_{Inc} at each RTCP interval instead of each RTT.

$$\beta = \frac{R_{Inc}}{t_{RTT}}, \text{ where } R_{Inc} = \frac{\text{RTCP interval}}{t_{RTT}} \tag{9}$$

If (Packet loss rate > 0):   // congestion-state
    If (Buffer occupancy < $MinTH$)
        Decrease video quality
    Else if ($MinTH$ < Buffer occupancy < $MaxTH$)
        Sending rate -= α
    Else (Buffer occupancy > $MaxTH$)
        Increase video quality
    Update $MinTH$ and $MaxTH$
Else (Packet loss rate = 0):   // stable-state
    If (Buffer occupancy < $MinTH$)
    Decrease video quality
    Else if ($MinTH$ < Buffer occupancy < $MaxTH$)
        Sending rate += β
    Else (Buffer occupancy > $MaxTH$)
        Increase video quality
    Update $MinTH$ and $MaxTH$

Fig. 5. MaVIS's algorithm

Figure 5 shows the pseudo code of our MaVIS mechanism. Depending on the network state and estimated buffer occupancy, we decide whether the sending rate or the video quality should be controlled. If the estimated receiver buffer occupancy is very high or low, then the MaVIS controls the video quality to prevent buffer underflow or overflow. Otherwise, the MaVIS controls the sending rate for the network stability according to the current network state. Since the estimation of receiver buffer occupancy is related to the encoding rate of video stream, two threshold values must be updated accordingly, if the video quality is changed.

4   Simulation and Evaluation

4.1   Simulation Environment

In this Section, we present our simulation results. Using the ns-2 simulator, the performance of the MaVIS mechanism has been measured, compared with the TFRC and the buffer-driven scheme [16]. To emulate the competing network conditions, background TCP Reno traffics were introduced.

Figure 6 shows the network topology for our simulations. We assume that the video quality is controllable by using the scalable video codec. Under this assumption, five different video qualities are used, such as 20Mbps, 16Mbps, 12Mbps, 8Mbps, and 4Mbps. 8Mbytes buffer space is allocated for the receiver buffer. We assume that the initial buffer occupancy is 50%.

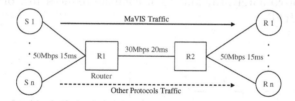

Fig. 6. Simulation environment

4.2   Performance Evaluation

Before discussing the performance of the MaVIS mechanism, we first examine the accuracy of the buffer estimation algorithm. Because the MaVIS controls either the sending rate or the video quality based on the estimated buffer occupancy, the accuracy of estimation algorithm significantly affects the performance of our mechanism. Figure 7 compares the estimated buffer occupancy in a sender side and the actual one in a receiver side. It shows that the estimated buffer occupancy and the actual one are very close. It is also shown in Fig. 7 that the media-aware thresholds are adaptively changed based on encoding rate of the video stream.

To evaluate our proposed mechanism, throughput, video quality, buffer occupancy, and packet losses are measured. Figure 8 (a) shows that the MaVIS dynamically controls the sending rate and the video quality. In the beginning, the sender transmits the video stream with 16Mbps. At about 21, 56, 80 and 102 second, the sender increases the video quality in order to prevent buffer overflow.

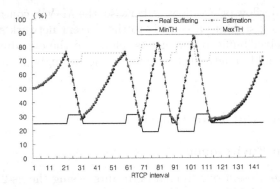

Fig. 7. Estimation of receiver buffer occupancy

At about 30, 68, 85, 91 and 112 second, the sender decreases the video quality in order to prevent buffer underflow. In these cases, the sender only controls the video quality without the sending rate control. Figure 8 (b) shows the packet losses of the MaVIS mechanism, compared with RTP. Unlike RTP, the MaVIS can reduce packet losses by controlling the sending rate based on current network state. Approximately, the MaVIS mechanism reduces 40% of packet losses, compared with RTP.

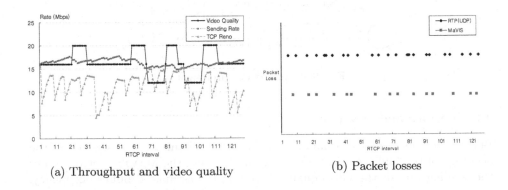

(a) Throughput and video quality

(b) Packet losses

Fig. 8. Performance of the MaVIS mechanism

In Fig. 9, the MaVIS is compared with the previous streaming protocols, the TFRC and the buffer-driven scheme in terms of the packet losses and the buffer occupancy. Figure 9 (a) shows the changes in the buffer occupancy. The TFRC experiences a serious buffer underflow, because it has no control on the video quality upon the buffer occupancy. However, the MaVIS mechanism and the buffer-driven scheme are successfully preventing the buffer underflow or overflow by adaptively controlling the video quality. In Fig. 9 (b), the packet losses for each protocol are shown. Because the buffer-driven scheme controls the video quality based on the current buffer state without the consideration of the network state, it suffers more packet losses than our proposed mechanism. However, the packet losses in the MaVIS and the TFRC are about the same. From this result, it is shown that the MaVIS mechanism has the approximately same performance with the TFRC in the aspects of the network stability.

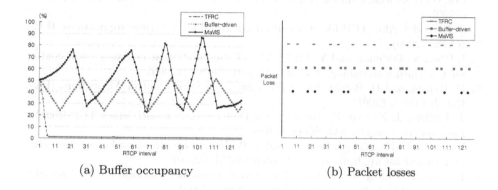

(a) Buffer occupancy   (b) Packet losses

Fig. 9. Comparison between MaVIS and previous works

5 Conclusion

In this paper, in order to overcome limitations of the previous streaming protocols, we propose a new mechanism for efficient video streaming, called MaVIS. The MaVIS mechanism considers both user and network requirements. It controls the sending rate or the video quality on the basis of both the estimated buffer occupancy and the network state. The MaVIS improves the network stability by reducing the packet losses and also provides the smoothed playback by preventing buffer underflow or overflow. Moreover, the MaVIS efficiently manages the buffer resource and reduces unnecessary sending rate and video quality changes because it is designed to consider the media characteristics of the video stream.

Simulation results have shown that our MaVIS mechanism has a better performance than previous approaches. In the future, we plan to apply the quality adaptation technique to our rate control scheme, as well as to perform real experiments in a real network environment.

Acknowledgement

This research was supported by the MIC(Ministry of Information and Communication), Korea, under the ITRC(Information Technology Research Center) support program supervised by the IITA(Institute of Information Technology Assessment).

References

1. S. Floyd and F. Kevin: Router mechanisms to support end-to-end congestion control. Technical Report, LBL-Berkeley. (1997)
2. S. Cen, C. Pu, and J. Walpole: Flow and congestion control for internet streaming applications. Multimedia Computing and Networking. (1998)
3. R. Rejaie, M. Handley, and D. Estrin: RAP: An end-to-end rate based congestion control mechanism for real-time streams in the Internet. IEEE INFOCOMM. (1999)
4. S. Na and J. Ahn: TCP-like flow control algorithm for real-time applications. IEEE ICON. (2000)
5. I. Rhee, V. Ozdemir, and Y. Yi: TEAR: TCP emulation at receivers - flow control for multimedia streaming. Technical Report, NCSU. (2000)
6. D. Bansal, and H. Balakrishnan: Binomial Congestion Control Algorithms. IEEE INFOCOMM. (2001)
7. J. Padhye, J. Kurose, D. Towsley, and R. Koodli: A model based TCP-friendly rate control protocol. NOSSDAV. (1999)
8. S. Floyd, M. Handley, J. Padhye, and J. Widmer: Equation-based congestion control for unicast applications. ACM SIGCOMM. (2000)
9. B. Song, K. Chung, and S. Rhee: Distributed Transport Platform for TCP-friendly streaming. LNCS 2662, Springer-Verlag Press. (2003)
10. S. Lee and K. Chung: TCP-friendly rate control scheme based on RTP. International Conference on Information Networking. (2006)
11. N. Aboobaker, D. Chanady, M. Gerla, and M. Sanadidi: Streaming media congestion control using bandwidth estimation. IFIP/IEEE Internation Conference on Management of Multimedia Networks and Services. (2002)
12. A. Balk, D. Maggiorini, M. Gerla, and M. Sanadidi: Adaptive MPEG-4 video streaming with bandwidth estimation. QoS-IP. (2003)
13. D. Ye, X. Wang, Z. Zhang, and Q. Wu: A buffer-driven approach to adaptively stream stored video over Internet. International Conference on High Speed Networks and Multimedia Communications. (2002)
14. J. Padhye, V. Firoiu, D. Towsley, and J. Kurpose: Modeling TCP throughput: A simple model and its empirical validation. ACM SIGCOMM. (1998)
15. H. Schulzrinne, S. Casner, R. Frederick, and V. Jacobson: RTP: A transport protocol for real-time applications. IETF, RFC 1889. (1996)
16. UCB LBNL VINT: Network Simulator ns (Version 2). http://www-mash.cs.berkeley.edu/ns/

On the Power of Cooperation in Multimedia Caching

Itai Dabran and Danny Raz

Computer Science Department, Technion, Haifa 32000, Israel

Abstract. Real time multimedia applications such as Internet TV, Video On Demand, Distance Learning, and Video Conferencing are becoming more and more popular over the Internet. Streaming media caching is a critical ingredient in the ability to provide scalable real-time service over the best effort Internet. In many cases, bandwidth becomes the system bottleneck and the cache cannot provide the required quality for all streams simultaneously. In this paper we study new algorithms, based on cooperation, which can improve the cache ability to provide service to all of its clients. The main idea is based on the willingness of streams to reduce their used bandwidth and allow other streams that may need it more, to use it. Our extensive simulation study indicates that our algorithms can reduce the pre-caching time by a factor of 3, or increase the probability for adequate service level by 30% using the same pre-caching time.

1 Introduction

A web proxy cache sits between web servers and clients, and stores frequently accessed web objects. The proxy receives requests from the clients and when possible, serves them using the stored objects. When considering streaming media, caching proxies become a critical component in providing scalable real-time services to the end users. Due to the bursty nature of streaming media amplified by compression, providing high quality service requires a considerable amount of bandwidth that may not be always available in the best effort Internet. The main problems that arise in this context are the large size of media items and their variable bit-rate nature: bandwidth consumed by the user is measured by frame per second (frame rate below 24 frames/second is distracting to the human eye) but the size of the frame is variable according to the multimedia visualization and compression method.

Figure 1 depicts an architecture of video delivery from video servers over the Internet to clients connected to a cluster of multimedia cache proxies, over a high speed LAN. The proxy may experience data losses, delays and jitter, while it has to provide the multimedia streams to its customers at high bandwidth and low delay. This is done, in many cases, by keeping a buffer containing several seconds of the stream content per each stream served by the cache. In times when the amount of data needed in the stream increases, or when the available bandwidth reduces due to competitive streams or other traffic load, the cache

A. Helmy et al. (Eds.): MMNS 2006, LNCS 4267, pp. 85–97, 2006.

still delivers the stream to the client, reducing temporary the amount of buffered data. In current streaming media cache proxies, each stream is handled indepen-

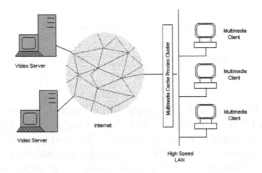

Fig. 1. Video Delivery Architecture

dently, as a separate object. That is, the server opens a TCP connection to the streaming media server, checks the buffer status and possibly other local parameters and then when needed, asks for more streaming data from the server. The downloading rate of the data depends, in addition to the cache requests, on the networking conditions (i.e. available bandwidth and load) and TCP behavior. When a cache proxies' cluster deals with more than one stream, the flows of the different streams may compete on the available bandwidth and they may experience losses. In such a case the TCP mechanism will reduce the flow rate and the amount of data in the flow's buffer may decrease below the ability to serve the client in the desired quality. Note that TCP is often used for media uploading both because unlike UDP it has a congestion control mechanism, and since it is much more resistant to Firewall related blockages.

The main novel idea of this work is to introduce cooperation between the flow management of different streams, all served by the same cache proxy or cache proxies' cluster. When congestion is detected, or when the amount of the buffer's data is decreased, the cache proxy can check the status of other flows and if their buffers are adequately full, reduce their flow by manipulating the TCP flow control. In this way, in scenarios similar to the architecture mentioned before, the load on the bottleneck link from the Internet to the cache proxy is reduced, enabling the needed flow rate to increase by using the TCP normal flow control mechanism. This approach is different from other schemes that are in use today and use the term cooperation. A classic caching cluster may hold multiple copies of the same movie, and cooperate in order to fetch the movie from a neighbor cache instead of fetching it from the server (using ICP [14] for example). On Video On Demand (VOD) systems when more than one client wants to watch the same movie in another time interval, cooperation is used in order to "patch" the needed multimedia stream for the client that joins later.

In this paper we use the general term Multimedia Caching Cooperative Device (MCCD) for a streaming media cache proxy or cache proxies' cluster. In our scheme, when the MCCD shares the same Internet connection, it can optimize the buffer usage in order to increase the scalability of the streaming media, by controling the rate of the incoming streams. The devices in a cluster can exchange information about the storage capacity (or a lack of input bandwidth) in order to enable one device to decrease its rate when it can, and by this to increase the bandwidth towards another one in order to overcome a buffer shortage. The same idea could be used in 3G/4G cellular networks where again, multiple streams compete on limited bandwidth. However in cellular networks, the bandwidth bottleneck is between the cache proxy and the clients that use the cellular devices. In these scenarios, achieving flow control cooperation is more difficult, yet the benefit of cooperation remains very high. We present two algorithms that manipulate the TCP flow control in order to reduce the load on the bottleneck link when needed. The first one, "Lower Threshold Algorithm" (LTA), reduces the TCP receiver window size when the cache buffer's size is decreased below some predefined threshold. The second, "Upper Threshold Algorithm" (UTA) reduces the TCP receiver window size of each connection when its cache buffer's size increases above a certain predefined threshold and by this smoothes the flow control of one recipient. Our extensive simulation study indicates that using our algorithms can reduce the pre-caching time by a factor of 3, or increase the probability for adequate service level by 30% using the same pre-caching time. We also address an implementation approach, an algorithm operated from the application layer, since the TCP Layer generally does not have an application interface for controlling the TCP flow or congestion control.

This paper is organized as follows. Section 2 describes related work used in multimedia caching in order to efficiently cache and use web items. Section 3 introduces the algorithms used in order to cooperate between the MCCD proxies. Section 4 shows the evaluation and the advantages of the proposed scheme, and Section 5 describes a practical approach that can be easily adopted in order to implement our idea. Finally, Section 6 presents our conclusions.

2 Related Work

In order to avoid the mass storage location needed in order to cache multimedia streaming objects, proxies use the fact that it is not necessary to cache all the streaming media. Streaming media nature allows them to retrieve new frames during the time that the user retrieves the cached frames. In [11], a "Prefix Caching" algorithm that caches an initial portion of the streaming media is proposed. The proxy can deliver this "prefix" to the client upon a request and meanwhile to cache the remaining portion of the media. The idea of "Prefix Caching" mechanism consists of caching a group of consecutive frame at the beginning of the stream in order to smooth and reduce the variable bit-rate of the streaming media. Another algorithm, the "Video Staging" algorithm is proposed in [15]. In this algorithm, the proxy caches a portion of bits from all the frames whose size is above some predefined threshold, and uses them in order to smooth bursts when such a frame is transmitted to the user. Efficient line bandwidth consumption is handled by several caching strategies designed for multimedia caching. Some partial caching strategies are described in [7] [8] and [15] whereas only certain parts of the multimedia stream are cached.

In order to enable a proxy to control the portion of the cached media the authors of [8] propose a "frame wise" selection algorithm in an architecture similar to the one mentioned in Section 1 whereas a proxy cluster is connected to a video server over the Internet, and to its clients over a high speed LAN. This algorithm, "Selective caching for Best effort Networks" (SCB), iteratively selects frames located between the closest buffer peak t_{max} and the risky time t_r afterwards and caches them according to the buffer-space constraints. This algorithm prevents troughs in the proxy buffers and has a better robustness than "Prefix Caching" methods under the same buffer space limitations. In [9, 10] it is proposed to cache more layers of popular videos and by this to overcome congestions that may appear later on the network. The authors of [12] propose few schemes aimed at minimizing the bandwidth needed from the origin server. Each of these schemes is applicable in different scenarios depending on the bandwidth, cache-space tradeoffs and service requirements. For example, partial caching of "Patch" and regular channel is done by re-use of buffers allocated by each "Patching-Window" interval for subsequent intervals.

Video On Demand (VOD) Multimedia stream can be received over multiple channels. A technique mentioned in [5] proposes to use two channels, the first for the complete streaming media, and the second for "patching" the data when a client joins later on to the streaming media. The client receives both channels, caches the data from the first channel, and uses it after playing the data received from the "patch" channel. In [5] it is shown that the "Patching" scheme supports much higher requests rate for VOD.

Cache clustering is a natural way to scale as traffic increases. Different schemes are used in order to arrange such a cluster, for example, in the *loosely-coupled* scheme proposed in [3], each proxy in the cluster is able to serve every request, independently of the other proxies. Popular protocols for this scheme are ICP (Internet Cache Protocol) [14] and HTCP (Hyper Text Caching Protocol) [13]. By using these protocols, proxies can share cacheable content, and increase the end user performance and the availability of bandwidth towards the server. In [1], "MiddleMan" - cooperating video proxy servers connected by a LAN, and organized by "Coordinators" is presented. A "Coordinator" process keeps track of the files hosted by each proxy and redirects requests accordingly. It was found that when more proxies are used, smaller peak loads in each proxy are created and better load-balancing between them is achieved. Since in a cluster, there is no need to cache the same streaming multimedia object in different copies for more than one user, the cluster should be able to maintain a management of a single copy inside it. Such a solution "The Last-Copy approach" for maintaining one copy of web items in a proxies' cluster in mentioned in [2], where an inter-cluster scheme is used in order to share the proxies' resources, and by this to provide scalable service to the cluster users.

3   Cooperative Protocols

Streaming Multimedia is used today for Internet-TV, Video On Demand, Distance Learning and many other applications. Suppose that MCCD (a cluster of proxies managed by a director for example) who serves several clients via a fast LAN, is connected to the Internet through a broadband connection. In order to overcome the delay and jitter caused by the Internet nature it may use any of the algorithms described before, such as the "Prefix Caching" algorithm [11], or the "Video Staging" algorithm [15]. Each Multimedia session is based on a TCP connection to a video server and has a predefined buffer space. Since the bandwidth of the connection from the MCCD to the Internet is bounded, we focus on balancing of the streaming media bandwidth between the different flows. Suppose a Home-MCCD retrieves several movies for family members, each from a different video server, via the Internet. If the buffer size of one of the proxies becomes empty, the movie transmission towards its specific user fails. Thus, when the content of a buffer decreases below a given threshold, more bandwidth is needed in order to keep the stream alive. Lowering the rate towards other video servers enables more available bandwidth and the TCP mechanism increases the rate, filling faster the buffer that is under risk. In order to implement such a cooperation, the MCCD needs to maintain a database of its buffer status for each multimedia stream. In this paper we focus in the case of a Home-MCCD combined of two proxies' processes, that retrieves two movies, m_i and m_j for family members. In order to rescue a movie whose buffer content is too small, we used two different algorithms that control the TCP receiver window. By reducing the receiver window of the other movie, the server is requested to reduce the input rate, enabling more bandwidth towards for the other connection. In the first algorithm "Lower Threshold Algorithm" (LTA), we check each time if the buffer size of each movie is below some predefined threshold. If it is, we reduce the TCP receiver window size of the second link to half of its value, and then adjust the threshold to half of its original value in order to keep decreasing the TCP receiver window if the buffer size continues to drop down. The pseudo

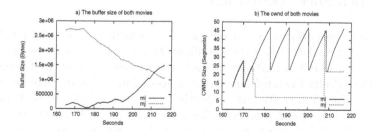

Fig. 2. The behavior of the LTA scheme

code of the "Lower Threshold" algorithm is presented in Algorithm 1. We keep on decreasing the receiver window of the other connection and the temporary threshold, until the problematic buffer recovers. When we recognize such a recovery (lines 24 or 31 of the algorithm) we reset all the parameters and go back to normal mode. Figure 2 shows how this algorithm behaves. Figure 2(a) depicts the amount of data in the buffers and Figure 2(b) depicts the behavior of the TCP cwnd, on the same time interval. Typically, before it drops, the cwnd of both movies is in the congestion control section of 15 to 28 segments. As can be seen when the buffer size of m_i drops below some threshold (and we can see that the buffer of m_j remains high) then the cwnd of m_j starts to be bounded by its receiver window, and drops down to the value of 7 segments. By this, the bandwidth towards the first connection increases, and its cwnd changes up to 45 segment, filling quick its buffer until it increases above the threshold. The threshold taken should be optimized according to the input bandwidth and to the caching algorithm in use. If enough bandwidth is available then the threshold should be high, since the buffer of one movie can be filled quick when we increase back its receiver window, and we can reduce the risk of a failure of the other one when its buffer will drop below zero. The caching algorithm also influence the decision about the threshold size. If the "Prefix-Caching" algorithm is used for example, then frame-bursts may appear later and the threshold should be high enough in order to prevent it. If a caching algorithm that smoothes the frame-bursts is used, then the threshold can be lower. However, when there is a limited amount of bandwidth, if the threshold is too low when both buffers are mostly empty and at least one of them is still below the threshold, then the other one may slow the connection too early without enough data to overcome this action. In contrary if the threshold is too high, then the algorithm may slow the connection of one movie too many times, increasing its failure risk. In the second algorithm "Upper Threshold Algorithm (UTA) we limit the buffers capacity by using a pre-defined upper threshold for the buffers. This prevents each of the connections to consume more bandwidth than needed. This solution has some disadvantage, since some of the available bandwidth may not be consumed. A pseudo code of this algorithm is presented in Algorithm 2.

4   Performance Evaluation

We evaluate the performance of our proposed scheme, by simulating it using NS (Network Simulator) [6] over a configuration consists of a home broadband network with an MCCD that uses two proxies' processes. In our first experiment we used the "Prefix Caching" algorithm proposed in [11] and in the second one a variant of the "Video Staging" algorithm proposed in [15]. In both cases we checked how each of our protocols, the UTA and the LTA, affects the MCCD success rate (i.e. full delivery of both movies). The trace we used was taken from [4]. It consists of 174,136 integers representing the number of bits per video frame over about two hours of the Star-Wars movie MPEG1 decoding. The inter-frame time of the StarWars trace is 1/24 sec. That means that the inter-arrival time for the arriving bits is 1/24 seconds. We split the frames of the first two hours (172,800 frames) into 8 short movies of 15 minutes each. Our experiments were

---

Algorithm 1. "Lower-Threshold Algorithm"

1: $pthold \leftarrow$ predefined threshold parameter
2: $ReceiverWindow_i \leftarrow$ the size of the the TCP receiver window of m_i
3: $ReceiverWindow_j \leftarrow$ the size of the the TCP receiver window of m_j
4: For each incoming frame do the following:
5: $s_{i(t)} \leftarrow$ the current size of the buffer of m_i at time t
6: $s_{j(t)} \leftarrow$ the current size of the buffer of m_j at time t
7: Normal-State:
8: **if** $s_{i(t)} < s_{j(t)}$ and $s_{i(t)} < pthold$ **then**
9:      $OldReceiverWindow_j \leftarrow ReceiverWindow_j$
10:     $ReceiverWindow_j \leftarrow ReceiverWindow_j/2$
11:     $thold_i \leftarrow pthold/2$
12:     Change into Recovery-State
13: **end if**
14: **if** $s_{j(t)} < s_{i(t)}$ and $s_{j(t)} < pthold$ **then**
15:     $OldReceiverWindow_i \leftarrow ReceiverWindow_i$
16:     $ReceiverWindow_i \leftarrow ReceiverWindow_i/2$
17:     $thold_j \leftarrow pthold/2$
18:     Change into Recovery-State
19: **end if**
20: Recovery-State:
21: **if** $s_{i(t)} < s_{j(t)}$ and $s_{i(t)} < thold_i$ **then**
22:     $ReceiverWindow_j \leftarrow ReceiverWindow_j/2$
23:     $thold_i \leftarrow thold_i/2$
24: **else if** $s_{i(t)} > pthold$ **then**
25:     $ReceiverWindow_j \leftarrow OldReceiverWindow_j$
26:     Change into Normal-State
27: **end if**
28: **if** $s_{j(t)} < s_{i(t)}$ and $s_{j(t)} < thold_j$ **then**
29:     $ReceiverWindow_i \leftarrow ReceiverWindow_i/2$
30:     $thold_j \leftarrow thold_j/2$
31: **else if** $s_{j(t)} > pthold$ **then**
32:     $ReceiverWindow_i \leftarrow OldReceiverWindow_i$
33:     Change into Normal-State
34: **end if**

---

done with all the possible 28 pairs that can be created from these 8 movies. These movies have a very bursty nature. The maximum frame size in this trace is 185267 bits, and the average one is 15611 bits. Hence, in order to to be able to pass the maximum frame (each $1/24seconds$), the needed bandwidth is 4.4Mbps, but the average bandwidth needed for all the movies is only 374Kbps, more than 12 times less than the maximum. In our most bursty movie derived from this trace the ratio between the peak bandwidth to the average one is 10.7, while in our moderate one, this ratio is 6.7.

We used a bandwidth interval between 600Kbps to 1.2Mbps for both MCCDs. The lower threshold used in our LTA protocol was 1.2M bytes when the "Prefix-Caching" algorithm was used, and $\alpha * Input-Bandwidth$ when the "Video-Staging" algorithm was used. We used $\alpha = 6.4$ (a threshold of 6.4Mbit for 1Mbps for example). We decreased the lower threshold in this case since the "Video-Staging" algorithm smooths and reduces the variable bit-rate of the future frame-bursts, thus less overhead should be taken. In our UTA experiments we used 2M bytes as our upper threshold. The "Prefix Caching" algorithm was implemented over the first 15 seconds, whereas the prefixes of both movies were cached before start consuming them from their buffers. The "Video Staging" algorithm was implemented for frames larger than 60K bits. The total average size of the portion of

---

Algorithm 2. "Upper-Threshold Algorithm"

1: $ReceiverWindow_i \leftarrow$ the size of the the TCP receiver window of m_i
2: $ReceiverWindow_j \leftarrow$ the size of the the TCP receiver window of m_j
3: $OldReceiverWindow_i \leftarrow ReceiverWindow_i$
4: $OldReceiverWindow_j \leftarrow ReceiverWindow_j$
5: $pthold \leftarrow predefinedthresholdparameter$
6: For each incoming frame do the following:
7: $s_{i(t)} \leftarrow$ the current size of the buffer of m_i at time t
8: $s_{j(t)} \leftarrow$ the current size of the buffer of m_j at time t
9: **if** $s_{i(t)} > pthold$ **then**
10:     $ReceiverWindow_i \leftarrow ReceiverWindow_i/2$
11: **else**
12:     $ReceiverWindow_i \leftarrow OldReceiverWindow_i$
13: **end if**
14: **if** $s_{j(t)} > pthold$ **then**
15:     $ReceiverWindow_j \leftarrow ReceiverWindow_j/2$
16: **else**
17:     $ReceiverWindow_j \leftarrow OldReceiverWindow_j$
18: **end if**

---

bits of frames larger than 60K bits, in our movies is 16Mbit. Thus, in order to pre-fetch it we need to consumes an average of 16 seconds for the staging phase in 1Mbps input rate. One condition was added to this algorithm: we assume that the cached "frame peaks" are accumulated in a different buffer. We check the influence of our schemes over the run-time buffer of the input stream.

We will use the terms m_a and m_b for the movies that are under consideration. On each case there is a "weak" movie m_a that may crash as a result of bursty frames interval, and a "strong" movie m_b that decreases its connection rate towards the server, in order to enable m_a to continue, using one of our schemes. We start with checking the overall validity of the scheme. Figure 3 shows the behavior of m_a and m_b when no algorithm is in use and when the LTA and the UTA algorithms are used. When no algorithm is used, after only 90 seconds the buffer of m_a gets empty, and the streaming media fails. However, when we use the LTA algorithm, the TCP receiver-window of m_b connection is decreased when the amount of data in m_a buffer drops below the predefined threshold. Then, the buffer of m_a does not get empty, and the streaming media continues with no interruptions. When the UTA algorithm is used, both m_a buffer and m_b buffer cannot increase beyond the upper threshold, and when one of them does, its TCP receiver-window is decreased in order to balance the transfer rate of the shared connection between the two connections. Again, we see that m_a does not crash, simply because m_b is not using all the available bandwidth, and by this enable m_a to compensate the buffer temporary shortage. Figure 4 depicts the improvement of the success to view both m_a and m_b, when the LTA algorithm is operated. These results are obtained when using the "Prefix Caching" scheme and the "Video Staging" scheme. We can see for example, that when the bandwidth varies between the ranges of 750Kbps to 850Kbps the LTA algorithm enables between 25% to 57% of the movie pairs to succeed, while without it there is only 10% to 35% pairs success. When the bandwidth is above 1050Kbps using the LTA algorithm causes all movies to be successful, while without it only 75% of the pairs are successful.

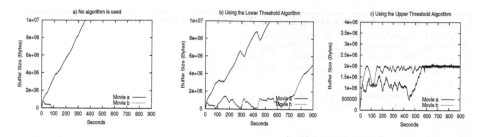

Fig. 3. Two movies behavior

Figure 4(b) shows that the "Video Staging" scheme improves the success of both movies more than the "Prefix Caching" one, but still, our LTA algorithm gets 20% to 30% better results than when no algorithm is used.

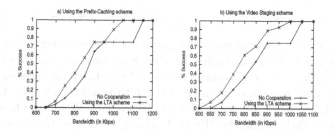

Fig. 4. The success of two movies when using the LTA Algorithm

5   A Practical Approach

Since the TCP Layer generally does not have an application interface for controlling the TCP flow or Congestion-control, implementing our algorithm in practice may be problematic. A possible solution is to temporary stop the TCP connection of one movie. However, this is not a smooth process, due to the TCP slow-start algorithm used in order to restart. Since bursts in a movie data do not last long, we can assume that time periods in which the buffer content decreases rapidly are short. We also assume that we should deal only with the case that during this time interval, only one movie is problematic, since if both of them are, we cannot do much. This discussion leads to the conclusion that stopping another TCP connection for a short time interval may solve the problem. When using a short time inteval we can use a "risky" lower threshold for m_a, lower than the one we used in the "Lower Threshold" algorithm. This reduces the possibility that m_b will crash also, since we fill its buffer more time. The "Disconnect by Lower Threshold Algorithm" (DLTA) is presented in Algorithm 3. Note that we never stop both connections, and the buffer size of the stopped connection can drop temporary below the predefined threshold.

Algorithm 3. "Disconnect by Lower Threshold Algorithm"

1: *pthold ← predefinedthresholdparameter*
2: For each incoming frame do the following:
3: $s_{i(t)}$ ← the current size of the buffer of m_i at time t
4: $s_{j(t)}$ ← the current size of the buffer of m_j at time t
5: **if** $s_{i(t)} < pthold$ **then**
6:   **if** m_i connection is active **then**
7:     stop the connection of m_j
8:   **end if**
9: **else**
10:   **if** m_j connection is stopped **then**
11:     start the connection of m_j
12:   **end if**
13: **end if**
14: **if** $s_{j(t)} < pthold$ **then**
15:   **if** m_j connection is active **then**
16:     stop the connection of m_i
17:   **end if**
18: **else**
19:   **if** m_i connection is stopped **then**
20:     start the connection of m_i
21:   **end if**
22: **end if**

We evaluated the advantages of the proposed scheme using the same simulation and input as before. The buffer size threshold we used was a function of the input bandwidth, the caching time (in the "Prefix Caching" scheme), and the caching scheme in use (since we need a lower threshold with the "Video Staging" scheme). For example, we used a threshold of 600K bytes when the input bandwidth was 1Mbps, and the "Prefix Caching" time was 15 seconds. Figure 5 depicts the success rate when the DLTA algorithm is used. Figure 5(a) results are obtained when the "Prefix Caching" scheme was used, and Figure 5(b) results are obtained using the "Video Staging" scheme. In figure 5(a) we can see for example, that when the marginal bandwidth for both movies is 800 Kbps, only in 21% of the cases we succeed to see both m_a and m_b when there is no cooperation, while when the DLTA algorithm is used, we more than double it to 53% of the time. The same can be seen in Figure 5(b), whereas the "Video Staging" scheme improve the success rate over the "Prefix Caching" algorithm, but our scheme still gets about 30% better. Figure 6 depicts the improvement

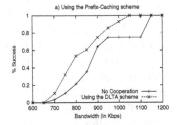
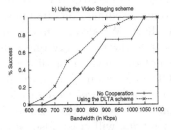

Fig. 5. The success of two movies when using the DLTA Algorithm

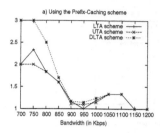
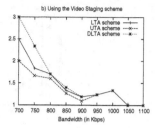

Fig. 6. The Improvement of the success of two movies

with respect to the no-cooperation case of all our proposed algorithms. We more than double the success rate when the link has a limited bandwidth, whereas in the lower bandwidth interval (below 800Kbps) the DLTA algorithm even triple this success. Another interesting result is the overall percentage of success to see movies with and without our scheme. As depicted in Figure 7, the DLTA schemes improves meaningfully the number of movies that can be seen via the MCCD for both caching algorithm while the other schemes do not. The main reason for this is that in a bandwidth shortage when in all other schemes both movies crash, the DLTA algorithm disables one of the connections, causes this movie to crash, but increases the bandwidth of the other, and let it to be finished.

A different way to evaluate our schemes is to compare the pre-caching time needed in order to be able to deliver both movies to the clients in all 28 cases. Figure 8 compares the needed pre-caching time in our schemes against the case where no cooperation is used. It can be seen that the caching time needed in order to achieve 100% success is 90 seconds when using input bandwidth of 1Mbps, and it is decreased to 5 second when the input bandwidth is 1200Kbps. However, when using our LTA scheme, it takes only 35 seconds, less than 34%, when the input bandwidth is 1Mbps. The DLTA scheme also shows a great improvement whereas the pre-caching time that guaranties success when the input bandwidth is 1Mbps is only 40 seconds and it drops to 15 seconds instead of 60 seconds when the input bandwidth is 1050Kbps.

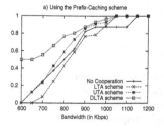
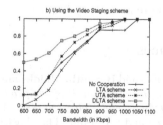

Fig. 7. The Improvement of movies success

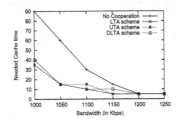

Fig. 8. Pre-Caching time in order to achieve 100% success for both movies

6   Conclusions

In this paper we proposed an efficient scheme for decreasing the pre-caching time, or increasing the probability for adequate service level of multiple streams using the same cache proxy. The heart of the proposed scheme, is an algorithm that utilizes cooperation between the flow management of different streams, which can improve the ability of the cache to provide service to all its clients. The idea is based on the willingness of streams to reduce their used bandwidth and allow other streams that may need it more to use it. In such a case, when congestion is detected, or when the amount of the buffer's data is decreased, the cache proxy can reduce other flows by manipulating the TCP flow control. Then, the load on the bottleneck link is reduced and the needed flow rate increases (or at least is not reduced), by using the TCP normal flow and congestion control mechanism. We also investigated a more practical approach that blocks the TCP connection for a short time interval, and showed that it may solve the problem. Our simulation results indicate that the proposed schemes achieve much better success rates (i.e. full delivery of both movies) than schemes in which no cooperation is used. Using our algorithms can reduce the pre-caching time by a factor of 3, or increase the probability for adequate service level by 30% using the same pre-caching time. A similar idea can be used in 3G/4G cellular networks where multiple streams compete on a limited bandwidth. In such networks, the bandwidth bottleneck is between the cache proxy and the clients that use the cellular devices, thus achieving flow control cooperation is a bit more complex, yet the benefit of cooperation remains very high.

References

1. S. Acharya and B. Smith. Middleman : A video caching proxy server. In *Proceedings of 10th International Workshop on Network and Operating System Support for Digital Audio and Video (NOSSDAV)*, June 2000.
2. R. Cohen and I. Dabran. The "Last-Copy" Approach for Distributed Cache Pruning in a Cluster of HTTP Proxies. In *7'th International Workshop for High-Speed Networks (PfHSN 2002)*, Apr. 2002.
3. I. Cooper, I. Melve, and G. Tomlinson. Internet Web Replication and Caching Taxonomy. RFC-3040, Jan. 2001.

4. M. W. Garrett and A. Fernandez. Variable bit rate video bandwidth trace using mpeg code. Available at: thumper.bellcore.com/pub/vbr.video.trace/ MPEG.description, Nov. 1994.

5. K. Hua, Y. Cai, and S. Sheu. Patching: A multicast technique for true video-on-demand services. In *ACM Multimedia*, pages 191–200, Sept. 1998.

6. S. McCanne and S. Floyd. ns-LBL Network Simulator. Available at: http://www-nrg.ee.lbnl.gov/ns/.

7. Z. Miao and A. Ortega. Proxy caching for efficient video services over the Internet. In *9th International Packet Video Workshop*, Apr. 1999.

8. Z. Miao and A. Ortega. Scalable proxy caching of video under storage constraints. IEEE J. Selected Areas in Communications, 20(7):1315– 1327, Special issue on Internet Proxy Services., Sept. 2002.

9. R. Rejaie, M. Handley, and D. Estrin. Quality adaptation for congestion controlled video playback over the internet. In *SIGCOMM*, pages 189–200, 1999.

10. R. Rejaie, H. Yu, M. Handley, and D. Estrin. Multimedia proxy caching mechanism for quality adaptive streaming applications in the internet. In *IEEE INFOCOM*, pages 980–989, Mar. 2000.

11. S. Sen, J. Rexford, and D. F. Towsley. Proxy prefix caching for multimedia streams. In *IEEE INFOCOM*, pages 1310–1319, Mar. 1999.

12. O. Verscheure, P. Frossard, and J.-Y. L. Boudec. Joint smoothing and source rate selection for guaranteed service networks. In *IEEE INFOCOM*, pages 613–620, Apr. 2001.

13. P. Vixie and D. Wessels. Hyper Text Caching Protocol (HTCP/0.0). RFC-2756, Jan. 2000.

14. D. Wessels and K. Claffy. Internet Cache Protocol (ICP). RFC-2186, Sept. 1997.

15. Z.-L. Zhang, Y. Wang, D. H. C. Du, and D. Shu. Video staging: a proxy-server-based approach to end-to-end video delivery over wide-area networks. *IEEE/ACM Transactions on Networking*, 8(4):429–442, 2000.

Delay Constrained Spatio-temporal Video Rate Control for Time-Varying Rate Channels

Myeong-jin Lee[1] and Dong-jun Lee[2]

[1] Dept. of Electrical Engineering, Kyungsung University, Busan, 608-736, Korea
mjlee@ieee.org
[2] School of Electronics, Telecomm. and Computer Engineering, Hankuk Aviation
University, Gyeonggi, 412-791, Korea

Abstract. In this paper, we propose a delay constrained spatio-temporal
video rate control method for lossy channels where the effective channel
rate available to the video encoder is time-varying. Target bit-rate constraint for encoding is derived, which guarantees in-time delivery of video
frames. By using empirically obtained rate-quantization and distortion-
quantization relations of video, the distortions of skipped and coded
frames in near future can be calculated in real-time. For the window expanding from the current to the firstly coded frame including the skipped
frames in between, the number of frames to skip from the current and the
quantization parameter(QP) for the firstly coded frame are decided in
the direction to minimize the average distortion of frames in the window.
From the simulation results, the proposed method is shown to enhance
the average PSNR performance compared to TMN8 with some increase
in the number of skipped frames and less number of delay violations.

1 Introduction

Most recently, with the increasing demand of video services over the Internet and
the wireless networks, adaptive video transmission over the lossy channels has
been the main focus of the research. Because there exists inevitable packet losses
and bit errors in the channels, video encoders adopt error protection, recovery,
and concealment mechanisms, which may require overhead at the expense of
some quality degradation. Packet-loss and bit-error ratios generally vary over
time, and they cause the effective channel rate available to the video encoder to
be time-varying. Thus, video encoders should adjust encoding parameters under
the end-to-end delay constraint continuously sensing the time-varying characteristics of channels.

Rate control plays an important role in the video encoder, which may have
great effect on the channel adaptability and the perceived quality. There have
been many research works on the rate control under the fixed frame rate or spatial resolution. However, because the coding complexity of video source and the
effective channel rate available to the encoder are time-varying, the overall distortion of video cannot be minimized by controlling just one encoding parameter,
i.e. a QP or a frame rate.

A. Helmy et al. (Eds.): MMNS 2006, LNCS 4267, pp. 98–109, 2006.

Recently, rate control algorithms[5, 6, 7] considered both temporal and spatial qualities jointly. However, additional buffering delay for pre-analysis of video source or the method of dynamic programming for optimal solving[6, 7] would not be applicable to real-time applications such as video phones and video conferences. Though the spatio-temporal optimization problem was simplified in [5] by using the explicit distortion models for coded and skipped frames, the distortion for skipped frames was not accurate enough and the end-to-end delay constraint was not considered.

In this paper, we propose a delay constrained spatio-temporal video rate control method which enables video encoders to efficiently adapt to the time-varying effective channel rate. In section 2, we discuss the delay constraint in video transmission systems and derive a constraint on the target bit-rate for encoding. In section 3, by using empirically obtained rate-quantization and quantization-distortion relations of video, the distortion models for skipped and coded frames in near future are proposed. In section 4 and 5, a problem is formulated and a real-time algorithm is presented for delay constrained spatio-temporal video rate control. For the window expanding from the current to the firstly coded frame including the skipped frames in between, the number of frames to skip from the current and the QP for the firstly coded frame are decided in the direction to minimize the average distortion of frames in the window. Simulation results and conclusion are presented in section 6 and 7, respectively.

2 Delay Constraint in Video Transmission Systems

For lossy channels, as shown in Fig. 1(a), joint source and channel coding is generally used to minimize the overall distortion by controlling source and channel coding parameters[3]. The effective channel rate available to the video encoder is time-varying because the rate allocation for the channel coding is done based on the time-varying characteristics of bit errors or packet losses. In this paper, we focus on the spatio-temporal video rate control method which can adapt to the time-varying effective channel rate. Joint optimization of source and channel coding parameters is not considered and left for further study.

Fig. 1(b) shows the video transmission system considered for spatio-temporal rate control. The encoder buffer is served with the effective channel rate mentioned above. We do not directly consider the time-varying characteristics of the channel, but only the time-varying effective channel rate available to the encoder.

Then, the encoder buffer occupancy is given by

$$B^e(j) = \max \left\{ B^e(j-1) + e(j, q_j) - s(j), 0 \right\}, \tag{1}$$

where $B^e(j)$, $e(j, q_j)$, and $s(j)$ are the encoder buffer occupancy, the generated bit-rate for the jth frame with the QP q_j, and the effective channel rate. The encoder buffer size is assumed to be sufficiently large.

The actual encoder buffer service rate is given by

$$\tilde{s}(j) = \min \left\{ s(j), B^e(j-1) + e(j, q_j) \right\}. \tag{2}$$

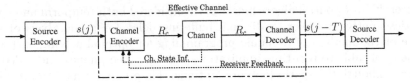

(a) video transmission system for lossy channels

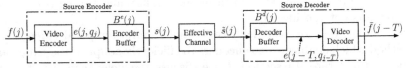

(b) video transmission system: source encoder's side view

Fig. 1. Video communication system

Then, the decoder buffer occupancy is given by

$$B^d(j) = \begin{cases} \sum_{i=1}^{j} \tilde{s}(i) - \sum_{i=1}^{j-T} e(i, q_i), & \text{if } j \geq T \\ \sum_{i=1}^{j} \tilde{s}(i), & \text{if } j < T \end{cases} \quad (3)$$

where T is the end-to-end delay between the input instances to the encoder buffer and to the decoder. For $j \geq T$, the decoder buffer occupancy can be modified as

$$B^d(j) = \sum_{i=j-T+1}^{j} \tilde{s}(i) - B^e(j - T), \quad (4)$$

where the encoder buffer occupancy is also represented by $B^e(j) = \sum_{i=1}^{j} \{e(i, q_i) - \tilde{s}(i)\}$ using Eq. 1 and 2. By applying the decoder buffer underflow condition $B^d(j) \geq 0$ to Eq. 4, we obtain the condition of $B^e(j) \leq \sum_{i=j+1}^{j+T} \tilde{s}(i)$. For buffered data, because the encoder transmits data in its maximum channel rate $s(j)$, we obtain $\tilde{s}(j) = s(j)$. Finally, the decoder buffer underflow constraint is given by

$$B^e(j) \leq \sum_{i=j+1}^{j+T} s(i), \quad (5)$$

for $j \geq T$. Eq. 5 means that the last frame $e(j, q_j)$ entering the encoder buffer during the jth frame period should leave the encoder buffer no later than the $(j + T)^{th}$ frame time.

For the case of the bounded network delay, the time left for the encoded data is decreased by the amount of the maximum transfer delay. Then, by combining Eq. 1 and 5, the constraint on the target bit-rate for encoding is given by

$$e(j, q_j) \leq \sum_{i=j}^{j+T_e} s(i) - B^e(j - 1). \quad (6)$$

where T_e is the time limit of video frames allowed to stay in the encoder buffer.

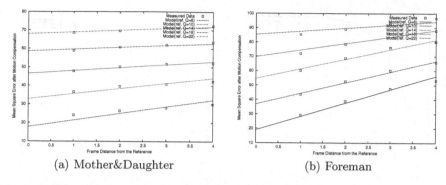

(a) Mother&Daughter   (b) Foreman

Fig. 2. Variance of the residual image after motion compensation($\sigma^2_{MC,j}$). The 100th frame is selected as the reference. QP's of 6, 10, 14, 18, and 22 are used for the reference.

3 Distortion Models for Coded and Skipped Frames

3.1 Distortion Model for Coded Frames

The distortion of the coded frame generally depends on the QP of the reference and the distance from it. Generally, QP's vary over macroblocks in a frame depending on rate control algorithms. But, it is difficult to find distortion models using the average QP because the distortion or generated bit-rate may also differ for the same average QP. Thus, in this paper, all the macroblocks in a frame are assumed to be quantized with the same QP. The distortion generally increases as the QP increases. As the distance from the reference increases, the variance of the residual image after motion compensation also increases. It is because the prediction efficiency gets lower as the distance from the reference increases.

The variance of the residual image depends on the coded distortion of the reference and the distance from it. Thus, the variance of the residual image for the jth frame, which is predicted from the j_c^{th} frame, could be modeled as

$$\sigma^2_{MC,j} = D_c(q_{j_c}) + \alpha \cdot (j - j_c), \tag{7}$$

where $D_c(q_{j_c})$ and α are the coded distortion of the reference and a constant parameter. An index j_c and a QP q_{j_c} are used for the reference frame.

Fig. 2 shows the relation of $\sigma^2_{MC,j}$ with the coded distortion of the reference and the distance from it. Though the proposed model is simple, it fits well with the measured data. The model can be used to predict the variance of the residual image without motion compensation and the result can be used to estimate the coded distortion of future frames without encoding.

To decide whether to encode or to skip a frame, the distortion model for coded frames should be investigated. There have been so many studies on the rate distortion relation for video coding. For real-time video transmission systems such as videophones, it is needed to estimate the coded distortion of incoming frames and to decide coding parameters fast, i.e. QP's, frame skip, and etc., in the

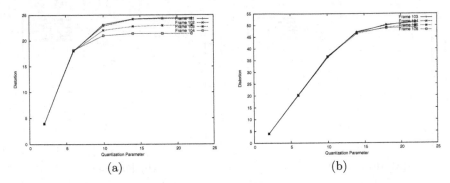

Fig. 3. Distortion of coded frames over QP's. *Mother&Daughter*. (a) Distance from the ref.=1, QP of the ref.=6 (b) Distance from the ref.=3, QP of the ref.=14.

direction to maximize the decoded video quality. Also, for wireless applications, the computational cost should be kept low for the longer battery life of mobile devices. Thus, it is needed to estimate the coded distortion fast enough with less computation.

Fig. 3 shows the distortion of coded frames over QP's for different combinations of reference distortions and the distances from the refence. While the coded distortion is linear for QP's below a threshold, it saturates to the variance of the residual image for QP's above the threshold. The threshold value generally depends on the QP of the reference.

Then, the distortion of coded frames can be estimated by

$$D_c(q_j) = \min\{c_j \cdot q_j, \sigma^2_{MC,j}\}, \tag{8}$$

where q_j and c_j are the QP and the rate-distortion parameter for the jth frame, respectively. The effect of the distance on the distortion is already considered in the variance model of the residual image.

For bit-rate estimation, the quadratic rate-quantization model[4] is used.

$$e(q_j) = a_j \cdot q_j^{-1} + b_j \cdot q_j^{-2} \tag{9}$$

where a_j and b_j are the model parameters.

For real-time applications, it is hard to get two rate-quantization points for parameter calculation. Though two rate-quantization points from adjacent frames could be used for the calculation, the difference between the QP's is not large enough for the model to cover the wider range of QP's. Also, changes in the characteristics of video source sometimes makes it difficult to get correct parameters.

By assuming a virtual rate-quantization point, q_v and $e(q_v)$, the model parameters can be calculated by using a single set of the average QP and the generated bit-rate of the recently coded frame. For real-time applications, the virtual QP q_v is set to a value outside the real range of QP's and the corresponding bit-rate $e(q_v)$ is set to a value less than the possible minimum bit-rate of frames.

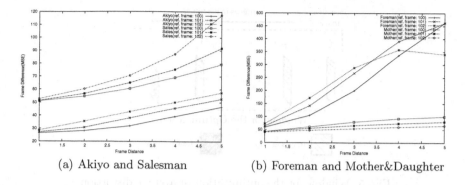

(a) Akiyo and Salesman   (b) Foreman and Mother&Daughter

Fig. 4. MSE of the source frame difference. Reference frame(j_c): $100 \sim 102^{th}$ frame.

3.2 Distortion Model for Skipped Frames

For skipped frames, if there is no frame interpolation or no other quality enhancing techniques used, the recently decoded frame is shown at their decoding times. Vetro analyzed the distortion for future skipped frames in [5] and it is given by

$$D_s(j_c, j) = D_c(q_{j_c}) + E\{\Delta^2 z_{j_c,j}\} \tag{10}$$

where j_c and j are the indexes of the recently coded frame and the future skipped frame, respectively. $\Delta z_{j_c,j}$ represents the source frame difference between the j_c^{th} and the jth frames.

To calculate the source frame difference, it is needed to predict the future skipped frame correctly. In [5], Vetro assumed that all pixels in the skipped frame have corresponding motion vectors and the motion vectors between frames are linear. Based on the assumption, the pixels in the skipped frame is predicted from the recently coded frame. However, it cannot be applied to real applications because the motion is not linear and the motion vector range is sometimes confined to provide error resilience. Also, the computational cost is high because the gradient and the motion vector should be calculated for all pixels.

Fig. 4 shows the mean square error(MSE) of the source frame difference with respect to the different frame distances. For neighboring frames, we argue that it is not the position of the reference frame, but the frame distance to greatly affect the MSE of the source frame difference. Thus, the source frame difference between the lastly coded frame and the future skipped frame can be predicted using the difference between the previous frames of the same distance. Then, the distortion for future skipped frames can be given by

$$D_s(j_c, j) = D_c(q_{j_c}) + f_d(j - j_c) \tag{11}$$

where $f_d(i)$ is the MSE of the frame difference between frames of distance i.

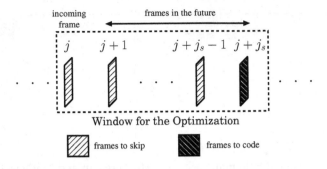

incoming
frame

frames in the future

j $\quad$ $j+1$ $\qquad$ $j+j_s-1$ $j+j_s$

Window for the Optimization

▨ frames to skip $\qquad$ ▧ frames to code

Fig. 5. Window for the optimization of average distortion

4 Problem Formulation for Delay Constrained Spatio-temporal Video Rate Control

Delay constrained spatio-temporal video rate control can be defined as the optimization problem of the average distortion under the delay constraint. The optimization is performed for the frame window shown in Fig. 5, which consists of successive skipped frames from the current and the next coded frame. Then, the problem can be formulated as follows.

$$\min_{j_s, q_{j+j_s}} D_{avg}(j_s, q_{j+j_s}) \tag{12}$$

$$s.t. \ e(j+j_s, q_{j+j_s}) \le \sum_{i=j}^{j+j_s+T_e} s(i) - B^e(j-1)$$

where j_s is the number of frames to skip from the current frame j.

An encoder distortion which includes the distortion for coded and skipped frames can be defined as follows.

$$D_{enc}(j) = \begin{cases} D_c(q_j), & \text{for encoded frames} \\ D_s(j_c, j), & \text{for skipped frames.} \end{cases} \tag{13}$$

Then, the average distortion over the frame window is given by

$$D_{avg}(j_s, q_{j+j_s}) = \frac{1}{j_s+1} \sum_{i=j}^{j+j_s} D_{enc}(i). \tag{14}$$

Under the assumption of stationary video source, by combining Eq. 10, 13, and 14, the average distortion can be calculated as follows.

$$D_{avg}(j_s, q_{j+j_s}) = D_c(q_{j_c}) + \frac{D_c(q_{j+j_s}) + \sum_{i=j}^{j+j_s-1} f_d(i-j_c)}{j_s+1} \tag{15}$$

5   Delay Constrained Spatio-temporal Video Rate Control Algorithm

5.1   Frame-Level Rate Control

Given the ranges of the number of frames to skip and the QP, the average distortion is calculated using the distortion models in Section 3. The rate controller determines a set of parameters and the corresponding target bit-rate which minimizes the average distortion.

Because the proposed distortion and rate-quantization models are based on the previous results of encoding and frame analysis, it is needed to update the model parameters frequently. Also, due to the time-varying characteristics of video source, the decided number of frames to skip(j_s^{min}) would not be accurate enough. Thus, only the current frame is skipped if j_s^{min} is larger than zero. Otherwise, it is encoded with the QP $q_{j_s}^{min}$. The frame-level rate control is performed in frame-by-frame basis.

5.2   Macroblock-Level Rate Control

If the current frame is encoded, the rate controller decides a QP for each macroblock. Basically, the QP decided in the proposed frame-level rate control is used as a lower bound for the quantization. Considering the time-varying characteristics of video, it is needed to regulate the bit-rate to meet the target. For this purpose, the QP in TMN8(QP_i^{TMN8}) is used as a reference and it is further adjusted considering the encoder buffer occupancy.

The encoder buffer occupancy is updated in macroblock level with the drain rate of the average bit-rate for a macroblock. For the jth frame, the encoder buffer occupancy after encoding the ith macroblock is given by

$$B_{MB}^e(i) = \max\{0, B_{MB}^e(i-1) + e_{MB}(i) - \frac{s(j)}{N}\} \tag{16}$$

where $e_{MB}(i)$ and N are the generated bits for the ith macroblock and the number of macroblocks in a frame, respectively, and $B_{MB}^e(0) = B^e(j-1)$.

The QP for the ith macroblock is given by

$$QP_i = \begin{cases} \max\{q_{j+j_s}^{min}, \min\{QP_i^{TMN8}, q_{j+j_s}^{min} + q_\Delta\}\} & \text{, if } \frac{B_{MB}^e(i-1)}{s(j)} > (T_e - 2) \\ QP_i^{TMN8} & \text{, otherwise} \end{cases} \tag{17}$$

where q_Δ is 2, 4, or 6 for the cases of $(T_e - 2) < \frac{B_{MB}^e(i-1)}{s(j)} < (T_e - 1.5)$, $(T_e - 1.5) < \frac{B_{MB}^e(i-1)}{s(j)} < (T_e - 1)$, or $\frac{B_{MB}^e(i-1)}{s(j)} > (T_e - 1)$, respectively.

Algorithm. Delay Constrained Spatio-temporal Video Rate Control

– *Step 0:* **Initialization**
  • $B^e(0) = 0, j_c = 0, j_s^{min} = 0, q_{j+j_s}^{min} = 0, a_j = a_0, b_j = b_0, c_j = c_0$

- *Step 1:* **Frame-Level Rate Control**
  - Given ranges of $j_s(j_s = 0, \ldots, j_{max})$ and $q_{j+j_s}(|q_{j+j_s} - q_{j_c}| \leq \Delta q)$
    * calculate the estimated bit-rate $e(j + j_s, q_{j+j_s})$ using Eq. 9
    * if $e(q_{j+j_s})$ satisfies the delay constraint and the average distortion(D_{avg}) is less than D_{avg}^{min}, update the parameter set of $\{j_s^{min}, q_{j+j_s}^{min}, D_{avg}^{min}\}$
  - If j_s^{min} is larger than 1, skip current frame and go to *Step 3*
  - else, go to *Step 2*
- *Step 2:* **Macroblock-Level Rate Control**
  - Update the encoder buffer occupancy using Eq. 16 in MB-by-MB basis
  - Quantization parameter decision for each macroblock
    * Calculate new QP_i^{TMN8} for the ith MB according to TMN8
    * Adjust the QP according to the buffer occupancy(Eq. 17)
  - Encode the ith macroblock with QP_i
  - After encoding all the macroblocks in current frame, $j_c = j$ and goto *Step 3*
- *Step 3:* **Parameter Update**
  - Update the encoder buffer occupancy $B^e(j)$
  - Update encoding parameters and distortion model parameters: a_j, b_j, c_j, q_{j_c}
  - Frame index increment: $j = j + 1$
  - go to *Step 1*

6   Simulation Results

6.1   Simulation Environment

In this section, we present some experimental results that demonstrate the performance of the proposed spatio-temporal video rate control algorithm. The simulations are based on the H.263+ codec[1]. We used three video sequences(*Akiyo, Salesman, Mother&Daughter*), all in QCIF format with frame rate of 30 fps. To evaluate the performance for different channel rates and delay constraints, we consider various combinations of channel rates($s(j) \in \{32, 64, var\}$ kbps) and encoder buffering limits($T_e \in \{2, 3, 4, 5\}$). To show how the proposed algorithm responds to time-varying channel status, the channel rate in Fig. 7 is used for simulation and will be refered as '*var*'. The time-varying bit error ratio in wireless channels with constant bit-rate(CBR) may cause the effective channel rate available to the encoder to vary over time as in Fig.7.

The performance of the proposed algorithm is compared with that of the TMN8[2] rate control algorithm. For the performance comparison of the algorithms, we used two measures of video quality. The first is the average PSNR to evaluate the spatial quality of video. For skipped frames, the PSNR is calculated by considering the recently decoded frame as the decoded frame. The PSNR of the skipped frames may be increased if the frame interpolation is applied. The second is the number of decoder buffer underflows to check how many frames have arrived at the decoder in time. The maximum deviation of the QP(q_Δ) is set to 3 to prevent abrupt quality variation. The larger the values, the less number of skipped frames, the larger quality fluctuation, and the larger probability of decoder underflow are expected.

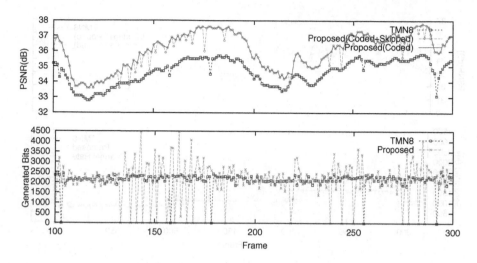

Fig. 6. Performance of the proposed algorithm. *Salesman*($T_e = 3$, $s(j) = 64$kbps).

6.2   Performance for the Constant Channel Rates

Fig. 6 plots the encoding results for *Salesman*. For the skipped frames, the generated bit is zero. The proposed algorithm shows enhanced PSNR performance with more frame skips over the sequence. There is somewhat larger PSNR fluctuation in the proposed, which is caused from the relative low PSNR of the skipped frames.

Table 1 shows the performance for *Akiyo* for various combinations of channel rates and the encoder buffering limits. As the channel rate or the encoder buffering limit increases, the average PSNR increases in the proposed algorithm. Because the TMN8 does not consider the end-to-end delay constraint for rate control, the average PSNR of TMN8 does not change for different encoder buffering limits. The number of skipped frames of the proposed is larger than TMN8. The number of decoder buffer underflows in TMN8 is generally larger than that in the proposed.

Compared with the TMN8, the proposed algorithm shows enhanced performance in the average PSNR and the number of decoder buffer underflows for different combinations of channel rates and encoder buffering limits. It is because the proposed algorithm determines the frame skip and the QP based on the distortion models for the coded and the skipped frames in the direction to minimize the average distortion within a window. The saved bits in the skipped frames are allocated to the next coded frames, which decreases the distortion of them. Then, the distortions of the following coded and skipped frames are recursively decreased because the distortion of the coded and skipped frames depends on that of the reference as in Eq. 7, 8, 9. The PSNR variation due to the skipped frames is somewhat large in the proposed algorithm. However, the PSNR of the skipped frames is generally larger than that of the corresponding frames in TMN8. Thus, we argue that the perceived quality of the proposed

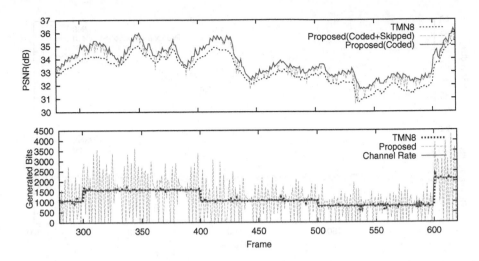

Fig. 7. Performance for the time-varying channel rate. *Mother&Daughter*($T_e = 3$).

Table 1. Performance of the proposed rate control method. *Akiyo*, 300 frames.

$s(j)$	T_e	Proposed			TMN8		
		skip	PSNR(dB)	dec. under.	skip	PSNR(dB)	dec. under.
32	2	65	35.4	0	7	35.5	7
32	3	63	36.3	0	7	35.5	6
32	4	67	36.2	0	7	35.5	5
64	2	52	38.6	0	4	39.0	4
64	3	40	39.7	0	4	39.0	2
64	4	45	39.8	0	4	39.0	1
var	2	61	37.9	0	4	38.3	7
var	3	48	39.0	0	4	38.3	3
var	4	51	39.1	1	4	38.3	1

algorithm is better than that of the TMN8. Also, the PSNR variation can be decreased by interpolating the skipped frames from the neighboring frames.

6.3   Performance for the Time-Varying Channel Rate

Fig. 7 shows the performance of the proposed algorithm for the time-varying effective channel rate. There are 0.4 dB enhancement over the TMN8 in the average PSNR for 900 frames. The number of the delay violations are 1 and 2 in the proposed and the TMN8 algorithms, respectively. There exist some delay violations in the proposed algorithm due to the abrupt changes in the channel rate. Because the number of delay violations depends on both the rate control method and the channel statistics, it is impossible to completely remove them. Thus, it is needed to jointly consider the source and the channel coding to minimize the overall distortion by source coding and channel errors, which is left for further study.

7   Conclusion

In this paper, we proposed a delay constrained spatio-temporal video rate control method. Target bit-rate constraint for encoding is derived, which guarantees in-time delivery of video frames. By using empirically obtained rate-quantization and quantization-distortion relations of video, the distortion models for skipped and coded frames in near future are proposed. The number of frames to skip from the current and the QP for the firstly coded frame are decided in the direction to minimize the average distortion of the frames. From the simulation results, it is shown that the proposed algorithm enhances the average PSNR performance compared to TMN8 with some increase in the number of skipped frames and less number of delay violations. The results can be utilized for video codecs based on motion-compensation and DCT, e.g. H.26X and MPEG-4, to adapt to the time-varying channel such as wireless networks and the Internet where the effective channel rate available to the encoder changes over time. For further works, it is needed to jointly control the spatio-temporal parameters and the code rate in channel codecs, which requires the video packetization, distortion and error propagation model for channel errors, and etc.

Acknowledgements

This research was supported by the Ministry of Information and Communication, Korea, under the Information Technology Research Center support program supervised by the Institute of Information Technology Assessment.

References

1. ITU-T Recommendation H.263, Version 2, 1998.
2. ITU-T/SG15, Video Codec Test Model, Near-Term, Version 8(TMN8), Portland, June 1997.
3. K. Stuhlmuller, N. Faber, M. Link, and B. Girod, *Analysis of Video Transmission over Lossy Channels*, IEEE J. Select. Areas Commun., Vol. 18, No. 6, pp. 1012-1032, 2000.
4. T. Chiang and Y.-Q. Zhang, *A New Rate Control Scheme Using Quadratic Rate Distortion Model*, IEEE Trans. Circuit Syst. Video Technol., Vol. 7, No. 1, pp. 246-250, 1997.
5. A. Vetro, Y. Wang, and H. Sun, *Estimating Distortion of Coded and Non-Coded Frames for Frameskip-Optimized Video Coding*, IEEE ICME, pp. 541-544, 2001.
6. S. Liu and C.-C. J. Kuo, *Joint Temporal-Spatial Bit Allocation for Video Coding with Dependency*, IEEE Trans. Circuit Syst. Video Technol., Vol. 15, No. 1, pp. 15-26, 2005.
7. R. C. Reed and J. S. Lim, *Optimal Multidimensional Bit-Rate Control for Video Communication*, IEEE Trans. Image Processing, Vol. 11, No. 8, pp. 873-885, 2002.

A Scalable Framework for Content Replication in Multicast-Based Content Distribution Networks

Yannis Matalas[1], Nikolaos D. Dragios[2], and George T. Karetsos[2]

[1] Digital Media & Internet Technologies Department,
Intracom Telecom, Athens, Greece
[2] School of Electrical and Computer Engineering,
National Technical University of Athens, Greece
imata@intracom.gr, ndragios@telecom.ntua.gr, karetsos@cs.ntua.gr

Abstract. This paper proposes a framework for replicating content in multicast-based CDNs. We focus on the design of a scalable and robust system that provides local availability and redundancy of content. The system takes on-line and distributed replication decisions on a per-object basis. The scalability and local redundancy is achieved by partitioning the overlay of surrogate servers into fully meshed groups. The proposed framework can incorporate any set of local metrics and constraints for deciding the placement of replicas, thus allowing the CDN designer to tune it to his specific deployment characteristics.

1 Introduction

Content Distribution Networks (CDNs) have become a common technology that enables content providers to distribute their popular content to a large number of users. Herein, we assume such a CDN system for distributing bulky files over a satellite network. Scalability, availability and efficiency of such a system are of vital importance especially in deployments that include a large number of receivers covering extended and distant regions. Efficient and scalable content distribution is achieved by applying: (a) multicast transmission, and (b) a distributed content replication algorithm that places content close to clients. The scalability of this algorithm is assisted by the partition of the CDN into relative small neighborhoods and the restriction of its scope within their bounds. At the same time, replication algorithm aims at providing content redundancy and load balancing in the CDN.

Different formulations for the problem of content placement in CDNs have been proposed in the literature, each one focusing on different objectives. From our perspective, content placement can be broken in two sub-problems. The first one, referred as *server placement* problem, is that of finding the locations where replica servers must be placed. *Server placement* is related to the design phase of a CDN and the deployment of the networking infrastructure. So far, it has been addressed in [2, 4, 9, 10, 11, 12] by algorithms, which are centralized, encompass high complexity and are executed off-line. The second sub-problem, referred as *content replication,* is the selection of the subset of the available servers to store replicas of a specific object in a way that minimizes the replication cost [3, 5, 7]. This is an optimization problem that must be usually solved on-line, i.e., the decision algorithm must run after some event. Hence, heuristics and distributed algorithms are preferable.

A. Helmy et al. (Eds.): MMNS 2006, LNCS 4267, pp. 110–115, 2006.

Herein, we address only *content replication*, and we propose a generic approach to its formulation and solution. Our objective is to describe a framework for this formulation, which permits an efficient on-line solution. This solution, is heuristic and sub-optimal, but has the advantage that is scalable and provides local redundancy guarantees for the content. Our approach has similarities with several previous approaches in its various aspects. It is distributed but considers cooperation of neighboring nodes as [5, 7, 8] have also suggested. It can take into account various different metrics, such as storage space availability, server load, previous user accesses and subscriptions [3, 5, 7]. On the other hand, it does not consider metrics related to delivery performance (e.g., latency) as in [2, 5, 6, 7, 9, 11]. In our view, network conditions should be considered later during request routing. Also, it can be assumed that statistical data about workload and link properties have been taken into account during a prior network design phase. Section 2 presents our CDN model and our content replication policy, and its evaluation is attempted in section 3.

2 CDN Model and Content Replication Policy

Satellite networks are a particularly appealing solution for content distribution and cache pre-filling due to their inherent broadcasting capabilities [1]. In our model, a one-hop satellite network is used for the distribution of large files from the origin server to a large number of geographically dispersed surrogate servers via IP multicast. Content distribution is triggered either by the content provider according to some schedule (push model), or by client requests (pull model). A few seconds before the eventual content transmission, the origin server multicasts an announcement in a well-known multicast address. All surrogate servers receive the multicast announcement, and each one decides whether it should replicate the advertised content or not. By applying a cooperative decision scheme, where all the servers in the neighborhood use the same decision policy based on identical neighborhood information, all servers eventually take a common decision. A surrogate that decides to replicate the content joins the corresponding multicast addresses (made known via the announcement) and receives it. A reliable multicast protocol, whose details are beyond the scope of this paper, is used for the content transmission to the surrogates.

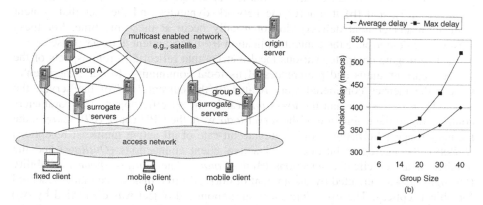

Fig. 1. (a) CDN model. (b) Simulation results for decision delay ($K=0.3*GroupSize$).

Fig. 1.a shows an abstract view of our CDN model that skips deployment specific details. A fully meshed grouping of the surrogate servers, based on geographical or network proximity criteria, has been determined during a prior network design phase. There are several reasons justifying this partitioning: (a) Replication decisions in a neighborhood must be independent from the decisions taken in other neighborhoods. (b) The server delivering to a client should not have long distance from this client. (c) The cooperation between servers in a group can ensure local content availability and redundancy. (d) The decision algorithm and the exchange of control messages inside a restricted neighborhood can be fast and incur only local traffic. (e) It maps to realistic CDN deployments where the nodes serving a geographical area (e.g., city, state, country) are relatively few and can be connected by fast links.

Ideally, if we replicated all objects in all the surrogate servers we would achieve the maximum possible content availability and the best delivery performance. There are, however, certain restrictions in doing so: (a) the overall storage space in a CDN is limited and, thus, we must restrict replication to the necessary, and (b) the distribution mechanism, although multicast-based, may present scalability issues due to the use of a reliable multicast protocol based on negative acknowledgements. A way to improve the CDN scalability is to multicast each object only in a subset of the receivers.

Apart from scalability, content replication aims at content availability in the sense of local replica redundancy and load balancing. In order to account for possible server failures and departures or conditions of overloaded servers, multiple replicas of an object should be available to support the demand in a given geographical area or in a given group of clients. To further enhance the content and system availability, the replication policy should avoid storing content in already loaded servers.

Most existing content placement approaches [2, 5, 6, 7, 9, 11] consider as known the locations of the clients and their distances from the candidate locations, and try to optimize the content delivery quality (e.g., minimize average latency). Using such a formulation they try to solve two problems in one step: the problem of finding the best server to replicate an object and the problem of finding the best server to deliver the object to clients. This formulation is valid for the off-line solution of the *server placement* problem where the distribution of clients and the network topology can be assumed to be static. In the case of *content replication*, however, such a formulation presents difficulties because: (a) its solution is computationally expensive and not scalable [6, 7], and (b) it ignores the network dynamics and the fact that content replication and content delivery do not usually occur at the same time. For these reasons, we de-correlate the content replication from the content delivery phase.

Each surrogate computes various local metrics that reflect different aspects of the server's current status and preferences of its local community. In our experimental system the metrics considered were related to the current *load* of the server, the available *storage space* and the *users interest* in the specific object. The parameters contributing to the estimation of the server *load* were the CPU usage (*CPUusage*), the memory usage (*RAMusage*), the aggregate bit-rate of all active multicast receptions (*InRate*), the aggregate bit-rate of all active client connections (*OutRate*) and the number of active client connections (*ActiveConns*). The *storage space* availability (*Storage$_{avail}$*) was reflected by the amount of currently free disk space that is reserved for object replicas. The *users interest* in an announced object was quantified by two terms: (a) the *Subscriptions* number that is the score found when we match the metadata of the object against the subscriptions of the local clients, and (b) the *PastUsage* that is the number of client hits to previous versions of the object.

When a surrogate server has computed the above values, it sends a message containing this value-set to all its neighbors. At the same time, all the neighbors perform exactly the same steps, and the result is that the value-sets describing the status of the servers are disseminated in the neighborhood. Each server finally collects an identical list of N value-sets (where N is the current size of the neighborhood), which then uses to evaluate a cost function and a number of constraints. A generic cost function for the replication of object i in server n could have the form:

$$replicationCost_{ni} = a_1f_1(CPUusage_n) + a_2f_2(RAMusage_n) + a_3f_3(InRate_n) + a_4f_4(OutRate_n) + a_5f_5(ActiveConns_n) + a_6f_6(Subscriptions_{ni}) + a_7f_7(PastUsage_{ni}) \qquad (1)$$

where $a_1,...,a_7$ are the weights of the various involved terms. An example constraint related to storage space is $Storage_{avail,n} > Storage_{thresh}$ (say $Storage_{thresh} = 100MB$). Of course, there is an infinite space of possible cost functions and constraints that could be alternatively applied. For instance, one could introduce a term related to storage space availability in the cost function above, e.g., $a_8f_8(Storage_{avail,n}, Storage_{max,n})$ or apply constraints related to server load such as (e.g., $InRate_n < 10Mbps$, $OutRate_n < 50Mbps$, $RAMusage_n < 80\%$). In any case, the focus of our work is not in the identification of a specific problem formulation, but in the design of a cooperative framework that permits the simple on-line solution of a whole family of formulations. Each surrogate server m computes the cost for all members in the group and checks if the identified constrains are satisfied. For the sub-list of group members that satisfy the constraints, the cost values are sorted and the members corresponding to the K lowest cost values are selected for local replication. If the server m is in the sub-list of the low cost servers it takes the decision to replicate the object. Note that exactly the same procedure is carried out in all servers, and since all servers of a group use identical input values, objective function and constraints, their decisions are identical.

There are different policies to define the number K of replicas taken in the group. A simple one is to set it proportional to the group size N, but not let it drop below a minimum value. This policy for K implies that all objects will have the same number of replicas in a group regardless from their popularity or their properties. Another option is to have K depending also on the *users interest* for the specific object. Also, we can have K depending on intrinsic object properties, such as its *importance* (e.g., the base layer of a scalable video is more important than the enhancement layers, and should have more replicas) and the *targeted audience* (e.g., when the content provider wants to increase or decrease the availability of specific objects in certain regions).

3 Evaluation

Scalability: An on-line content replication algorithm must have low complexity and should generate the least possible network traffic. As [6] suggests, existing centralized content placement algorithms [2, 5, 9, 10, 11, 12] are not scalable because the *computational complexity* increases with network size N (in the best case $O(N)$). In the proposed approach, the decision algorithm running in a specific node is independent from the state and the decisions taken at nodes belonging to other groups. Assuming a bounded group size (say $M \leq 20$), the computational complexity at any processing node is also bounded $O(M)$. Of course any distributed algorithm with restricted scope (neighborhood size) shares the same advantage. At the limiting end, purely local algorithms [3, 5, 7] have very low complexity $O(1)$ but the decision quality is lower. *Messages transfers* influence the scalability in two ways: They are additional network traffic and they add extra delay in the decision process. In a centralized algorithm each replication decision requires $2N$ message transfers. The problem with these messages is that many of them travel long distances through several network hops and through links that may be slow. Thus, the effective network traffic is higher and the added delay due to these messages is large and, in the general case, increases as the size of the CDN grows. In our algorithm, each surrogate sends $M-1$ messages to its neighbors and receives $M-1$ messages from them. Hence, the number of messages required to take replication decisions for an object in the M nodes of a group is $M(M-1)$, but these messages do not induce end-to-end traffic. Also, the average added delay due to message transfers can be always kept below some threshold if the network distance between group members is bounded. In general, the delay due to message transfers of distributed algorithms decreases when the scope of the algorithm becomes narrower. In fig. 1.b we have plotted the simulation results for the replication decision delay (mean values of average and maximum decision delay) for different group sizes in a 100Mbps LAN environment. These results show the applicability of our algorithm for on-line replication decisions, since for relatively small groups the replication decision delay is acceptable (<1 sec).

Local content redundancy: A drawback of existing centralized and purely local replication algorithms is that they do not provide any guarantees for local redundancy of replicas since they do not set any constraints related either to the relative placement of replicas or to the number of replicas taken in a specific region. Our approach inherently guarantees local redundancy as each group of surrogates takes exactly K replicas, where K may be derived in different ways (see previous section).

Flexibility of problem formulation and solution: The decision quality of most centralized algorithms is related to the client perceived latency. However, the solution ignores the actual network conditions, which may change dynamically and deviate a lot from the initial assumptions, and does not account for mobility of clients. Also, various assumptions are made in order to simplify the solution of the optimization problem, e.g., the constraints are relaxed and incorporated in the cost function [7,11], or large and small objects are not differentiated [7]. It is not always easy to incorporate additional metrics in their cost or constraint functions without serious impact in their complexity. On the other hand, our approach makes no simplification and the constraints are always satisfied by the solution. It is flexible and can easily incorporate any type of local metrics and constraints according to the model at hand.

Easiness of deployment: Centralized approaches assume that the central node knows the network topology and the latency to any node or client. Also, keeping this central node synchronized with the contents of all the nodes is a very difficult task. Achieving quick decisions and synchronization in such a system requires that the links with all the nodes are fast and reliable. The above requirements cannot be easily satisfied in real deployments with many nodes placed at distant locations. In our approach, configuration and synchronization involves only the neighborhood and is much simpler. Also, redundancy of processing nodes is not an issue. If any node fails the algorithm runs without problem in other nodes. The quality of the network links is important only between the nodes of the same group. And it is reasonable to assume that in a real deployment these nodes are placed topologically close to each other.

4   Conclusions and Future Work

We have proposed a scalable and flexible cooperative approach for content replication that solves several problems encountered by centralized and purely local algorithms. The grouping of surrogates provides a trade-off between scalability and local replica redundancy. Our approach lies somewhere in the middle between: (a) centralized algorithms, which take a near-optimal number of replicas but are not scalable and do not care about their relative placement, and (b) purely local algorithms, which are simple and scalable but the number of replicas taken may be far from the optimal while they make no provisions at all for their relative placement.

An important topic for future work is the storage management of surrogate servers, and particularly the use of cooperative replica removal and regeneration policies.

References

1. A. Armon and H. Levy. Cache satellite distribution systems: modeling and analysis. In Proceedings of IEEE INFOCOM, 2003.
2. Y. Chen, R. H. Katz, and J. D. Kubiatowicz. Dynamic replica placement for scalable content delivery. In Proceedings of 1st Int. Workshop on Peer-to-Peer Systems, 2002.
3. M. Chen, J. P. Singh and A. LaPaugh. Subscription-enhanced content delivery. In Proceedings of WCW'03, 2003.
4. C. Huang and T. Abdelzaher. Towards content distribution networks with latency guarantees. In Proceedings of 12th IWQoS, 2004.
5. J. Kangasharju, J. Roberts, and K. W. Ross. Object replication strategies in content distribution networks. In Proceedings of WCW'01, June 2001.
6. M. Karlsson, C. Karamanolis and M. Mahalingam. A framework for evaluating replica placement algorithms. Technical Report HPL-2002-219, HP Labs, 2003.
7. M. Karlsson and C. Karamanolis. Choosing replica placement heuristics for wide-area systems. In Proceedings of IEEE ICDCS, 2004.
8. M. Korupolu, G. Plaxton, and R. Rajaraman. Placement algorithms for hierarchical cooperative caching. Journal of Algorithms, January 2001.
9. S. Jamin, C. Jin, A. R. Kurc, D. Raz, and Y. Shavitt. Constrained mirror placement on the Internet. In Proceedings of IEEE INFOCOM, 2001.
10. B. Li, M. J. Golin, G. F. Italiano, X. Deng, and K. Sohraby. On the optimal placement of web proxies in the Internet. In Proceedings of IEEE INFOCOM, 1999.
11. L. Qiu, V. N. Padmanabhan and G. M. Voelker. On the placement of web server replicas. In Proceedings of IEEE INFOCOM, 2001.
12. P. Radoslavov, R. Govindan, and D. Estrin. Topology-informed internet replica placement. In Proceedings of WCW'01, June 2001.
13. H. Yu and A. Vahdat. Minimal replication cost for availability. In Proceedings of the 21st annual Symposium on Principles of Distributed Computing, 2002.

VidShare: A Management Platform for Peer-to-Peer Multimedia Asset Distribution Across Heterogeneous Access Networks with Intellectual Property Management

Tom Pfeifer, Paul Savage, Jonathan Brazil, and Barry Downes

Telecommunications Software & Systems Group [TSSG]
Waterford Institute of Technology [WIT] – Carriganore Campus
Waterford, Ireland
Phone: +353-51-30-2927
t.pfeifer@computer.org,
{psavage, jbrazil, bdownes}@tssg.org

Abstract. The VidShare project develops a management platform for Peer-to-Peer multimedia asset distribution across heterogeneous access networks, fixed and mobile, with Intellectual Property Management and Protection (IPMP). A key objective is to build an architecture that is interoperable with different content management and IPMP solutions. The project utilises MPEG REL as a rights expression language and works with the evolving MPEG-21 Multimedia Framework. The platform supports a number of business models including rights based personal multimedia sharing, distributed multimedia management for corporations, and the retailing of video content including the super distribution of content.

Keywords: multimedia in peer-to-peer networks, mobile multimedia, mobile network management, digital rights management, multimedia middleware and frameworks, content distribution networking.

1 Introduction

Currently, multimedia technology provides the different players in the multimedia value and delivery chain (from content creators to end-users) with access to information and services from almost anywhere at anytime through ubiquitous terminals and networks. However, no complete solutions exist that allow different communities, each with their own models, rules, procedures, interests and content formats, to interact efficiently in complex, automated e-commerce scenarios.

In addition most multimedia content is not governed today by Intellectual Property Management and Protection (IPMP) systems. Where they do exist they are rudimentary and do not interoperate. This creates a major barrier to rolling out complex multimedia services that require this interoperability and respect for Intellectual Property Rights (IPR), such as media sharing, retailing or superdistribution of video content.

The VidShare[1] platform identifies and defines the key elements needed to support these multimedia services and their delivery chain as described above, the relationships

---

[1] The VidShare project has been supported by Enterprise Ireland within the Commercialisation Fund – Technology Development Phase 2005, TD/2005/229, 2005 – 2006.

A. Helmy et al. (Eds.): MMNS 2006, LNCS 4267, pp. 116–123, 2006.

between and the operations supported by them. Thus the key objective of the VidShare project was to research and develop a peer-to-peer e-commerce platform, that enables the distribution of multi-media content (with a particular emphasis on video) across heterogeneous access / distribution networks (Internet, Mobile, Digital TV) with a management framework that enables interoperability of content and IPMP systems.

The key differentiator that VidShare has over other P2P networks is its focus on standards based multimedia interoperability and Intellectual Property (IP) management and protection. Participants in a P2P network can bring lots of legitimate value to both IP owners and users.

Within this paper, we discuss current limitations in Section 2. The platform design and architecture is introduced in Section 3, starting with a scenario based analysis, the architectural overview, and the embedding of IPMP and e-commerce functionality. Section 4 presents aspects of the implementations, and in Section 5 we summarise.

2 Key Issues and Existing Technical Limitations

When looking at the situation of how multimedia content is managed and protected today, the key issues and technical limitations of existing products, processes and methodologies today are:

1. Current content management systems do not communicate with each other; content cannot readily be identified across content management systems, or accessed in an easy-to-use, distributed fashion.
2. Most of the multimedia content existing today is governed by at best rudimentary IPMP systems.
3. No IPMP system has yet emerged as a de-facto standard. While various IPMP systems exist today, no management framework exists to allow for interoperation amongst such systems.
4. One problem for End Users interacting with content today is the lack of interoperability between IPMP systems.
5. Users currently need to have very explicit format and technology knowledge in order to utilise format conversion tools to use content on different devices.
6. Owners of rights in content require the freedom to exercise their rights by choosing channels and technologies (including IPMP Systems) through which to offer and make available their content.
7. There is a lack of standardised methods for monitoring and detecting infringements of rights.
8. Consumers of content may in some circumstances require the freedom to manage their privacy, which includes interacting with content without disclosing their identity to any other User in the value chain; Note that through new technologies (e.g., the Internet), End Users increasingly become owners of rights in content.
9. The differences between the national and regional legislations relating to IP law (and the lack of any universal legislation to protect rights holders) challenge existing IPMP systems to accommodate the evolution of global commerce in a digital environment.
10. The lack of integration of P2P networks with IPMP systems.

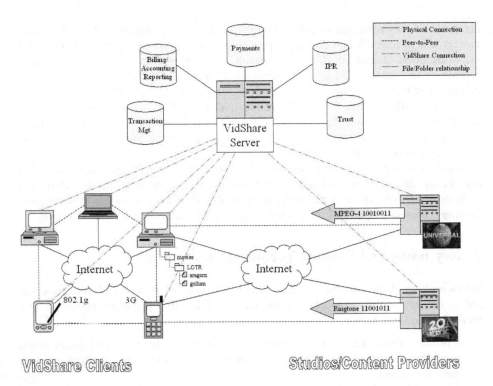

Fig. 1. VidShare high level architecture

3 Design and Architecture

3.1 Scenario Based Analysis

A key objective of the VidShare platform is to facility a number of innovative peer-to-peer based digital business models in a secure protected way. The *Scenario Based Approach* [1] has been used to understand the operation between owners providers and users of video content. The usage scenarios have been defined as follows:

1. *Rights based personal media sharing* – e.g. one user sharing personal video content, such as a family occasion, with the other family members in a secure, rights based manner (i.e. named users only can play the video, no editing or forwarding – a time limit of 1 month is specified).
2. *Rights based distributed corporate multimedia database* – for example the Dublin office creating and locally hosting content that the New York office can securely access and consume and visa versa. With rights based access, content can also be restricted by, for example, organization security clearances.
3. *Rights based e-tailing of video content.* In this scenario users will have the capability of searching for and then downloading and playing movies such as "The Day After Tomorrow". In this scenario the user has the right to play a trailer for the movie and then has the option to rent or buy the movie.

4. *The super distribution of video content.* In this scenario the content the user has downloaded has rights that enable the content to be forwarded to another user. A second user should be able to play the trailer and then have the option to rent or buy the movie. If the second user makes a purchasing decision the platform should have the flexibility to enable the original user to receive a percentage of the purchase price as a reward, in addition to settling with the players in the distribution chain.

3.2 Architecture

To enable the business models discussed in the scenarios above, the VidShare project implements the following technological components, as illustrated in Figure 1.

A distributed Peer-to-Peer architecture (P2P). This is a pure peer-to-peer network similar to the architecture of Gnutella for person-to-person and organisation-person sharing of secure rights based multimedia. This network requires augmentation with the VidShare management system to support the retailing and superdistribution of value based content to enable, content identification and IPR verification, the establishment of Trust relationships, value chain support, payments, accounting and settlement.

A client P2P application that supports:

1. User profile and preferences, content searching and downloading.
2. Functionality to enable a user to upload their personal video and encode/decode it with rights using the MPEG-21 REL standard.
3. Functionality to create a local, standards based, media database of sharable content for Peer-to-Peer sharing.

A management platform that oversees the Peer-to-Peer network which supports:

1. A framework to manage the P2P network including a register of peers.
2. A central index database for organisations and/or media companies where IPR validated content can be listed. Note: for legal reasons, private individual's content databases will be held locally and not in the central index database.
3. Value Chain Management and License services. This includes content License generation and distribution, content identification (and verification of IPR), content tracking and business model services such as superdistribution.
4. Trust Management Services. Registration of value chain participants, user certification and credential management services, trusted time services; in connection with Security and Protection Services. Secure channels, key management, trusted software and execution environment.
5. An e-commerce framework: Payments services, accounting services and settlement services; together with transaction, audit and log services. These services can be monitored by content providers and rights holders for auditing purposes.

3.3 IPMP Integration

VidShare utilises MPEG REL as a key element of its IPMP architecture, as MPEG is working with other organisations such as the OMA to integrate their approaches and ensure interoperability. Using the MPEG REL for content distribution provides the following benefits:

- It allows content distribution terms and conditions in precise XML-based language for both human and machine interpretation.
- It empowers content owners to manage the rights and conditions throughout the distribution channels, down to the level of content consumers, through the specification of the multi-tier distribution relationships, potentially in a single multi-tier MPEG REL license.
- It facilitates e-commerce by enabling the upstream content provider to present multiple offers (choices) to the downstream distributors or consumers, with the final selection presented as an enforceable license.
- It helps to automate the content distribution process by matching up the supply and demand entities based on contractual permissions and obligations.
- It enables the specification of any type of licenses, grants, rights, and conditions, by providing a comprehensive and extensible language. It provides unambiguous interpretation and interoperable expressions of MPEG REL licenses.
- It enables interoperability, and positions players in the content distribution value chain to use of best of breed components or web services and support interoperability across business models.

MPEG REL can be integrated directly into MPEG-4's binary format through MPEG-4's IPMP Hooks. The underlying philosophy is that the bitstream embeds information that informs the application which (of possibly multiple) IPMP system should be used to process governed objects in compliance with rules declared by the metadata. MPEG-4 integrates the hooks tightly with the MPEG-4 Systems layer, which makes it possible to build secure MPEG-4 delivery chains very efficiently.

There are two key extensions of basic MPEG-4 systems constructs:

- *IPMP-Descriptors (IPMP-Ds)* are part of the MPEG-4 object descriptors that describe how an object can be accessed and decoded. They are used to denote the IPMP System that was used to encrypt the object.
- *IPMP-Elementary Streams (IPMP-ES)* are all MPEG objects represented by elementary streams, which can reference each other. These special elementary streams can be used to convey IPMP specific data.

Using these tools, it is possible to have the IPMP architecture act on the media data; the scene description and the compositor; the interaction messages; and any other information conveyed in the BIFS (Binary Format for Scenes, MPEG-4's binary scene description / animation language).

3.4 E-Commerce and Management Capabilities

In addition to the IPMP architecture, VidShare integrates e-commerce capabilities into the management platform to support the project's business models. The key elements being a payments system, accounting system and settlement system to enable consumers to purchase and pay for content (and for content providers to receive the appropriate out-payments on sales of content).

As the key elements of this project relate to the P2P distribution of content with IPMP, exiting technology is integrated for the e-commerce framework rather than built for this project.

Thus the project integrates PayPal (www.paypal.com) for payments (which can work in a P2P environment) and the TSSG's Rating Bureau Service (RBS) for accounting and settlement.

4 Implementation

4.1 Components

The implementation of the VidShare system has been modularised so that custom configurations are possible for each target customer. Following is a detailed description of each module within the VidShare platform.

4.2 Mediation Component

The mediation component of the VidShare platform is responsible for the generation and hand-off of service usage records. These service usage records will detail each transaction authorised by the VidShare platform. A sample usage document is provided in Figure 2, adopting an XML format to keep the mark-up in a transportable and easily readable format.

The current mode of operation for the mediation component is to operate as the 'D' interface of the Network Data Management-Usage (NDM-U) specification [10]. This interface has been adopted as the immediate testing harness for the accounting aspect of the platform utilises the Rating Bureau Service (RBS), a TSSG rating service that exposes NDM-U interfaces and is IPDR compliant.

```
<?xml version="1.0"?>
<transaction type="authorisation">
        <content>VIDEO-MPEG-21</content>
        <contentId>3C5504E0-4FG9-18D3-CA0C-AB05E82C3301
        </contentId>
        <timeStamp>2006-05-04T14:08:34Z</timeStamp>
        <userId>vidshareUser@tssg.org</userId>
        <uniqueDeviceId>758493759584750</uniqueDeviceId>
</transaction>
```

Fig. 2. Sample instance document for a VidShare transaction

4.3 IPMP Component

The Intellectual Property Management and Protection (IPMP) component is responsible for interpreting the requests issued for authorisation of content by client devices. It is responsible for generating the correct licence, associated with the content being authorised. Initially this IPMP component has been developed to support the MPEG-21 Rights Expression Language (REL) [5]. However, it is viewed that this component will be enhanced to be compatible with the rights expression language of the Open Mobile Alliance (OMA).

4.4 Transaction Management

The transaction management component is currently a baseline implementation of transaction management functionality using the transaction management features of J2EE. As VidShare moves toward a commercial deployment it is viewed that this component will be replaced by a more substantial transaction manager (e.g. Tuxedo) as used by operators for their current customer services.

4.5 Trust Management

Trust management is critical to the successful operation of the VidShare platform. It is imperative that customers are assured of the security of their transactions; that content which they are acquiring is originating for a legitimate source and that any content or licence files distributed to them is verifiably safe and traceable. Without these measures in place, users of the system will not feel comfortable using content and licence providers that may be distributing malware or illegal content.

4.6 Payments Component

In order to fulfil the overall end-to-end provisioning of the service a basic payments solution has been implemented which uses the PayPal API to engage a micropayment solution for content e-tailing. PayPal serves as a useful mechanism for ensuring the security of users transactions and user experience by invoking a trusted third party payments processor. Again, it is envisaged that as VidShare moves toward a commercial deployment, this payment processor will be replaced by the in-house engines used by the mobile operators or other target customers.

4.7 VidShare Positioning

The VidShare platform is being positioned in the market as a plug-in solution for operators' networks. It will provide a standards compliant, black box interface for solving the DRM needs of network operators and the content providers who offer services over them. It is not viewed that VidShare will become a standalone service provisioning platform such as those currently seen in the Web 2.0 service domain (e.g. flickr.com) but rather it will be the behind the scenes enabler for mobile content distribution and super-distribution.

5 Summary and Future Work

VidShare is a Peer-to-Peer, e-commerce platform and client. It will enable content providers, such as 20th Century Fox, Universal, MTV, to distribute their video content to consumers in a secure environment to different types of devices such as PC's, mobile phones and evolving hybrid devices such as wireless multimedia terminals.

A key aspect of VidShare is that it offers the maximum distribution flexibility that a P2P network offers – but at the same time is controlled by a Intellectual Property Management and Protection (IPMP) system the supports the secure delivery of protected content, i.e. it prevents the content from being stolen or counterfeited. VidShare's management system also handles payment, transaction management, auditing, reporting, royalty and revenue settlement. VidShare in essence can automate a whole series of multimedia e-commerce business models.

Security issues are investigated in close cooperation with European Commission funded 6th Framework projects coordinated by the TSSG: SEINIT and SecurIST. We plan to fully exploit the results of the VidShare research by establishing a Campus Company which will revise the prototype as a commercial grade P2P system for marketing and distribution to the global marketplace.

References

[1] Kentaro Go; John M. Carroll: The blind men and the elephant: views of scenario-based system design. - interactions. ACM Press, New York, Vol. 11 (6) Nov+ Dec 2004, pp. 44-53

[2] ISO/IEC TR 21000-1, Vision, Technology and Strategy, July 2001 (DTR, N4333) and ISO/IEC PDTR 21000-1 Second Edition, Vision, Technology and Strategy, Dec 2003 (N6269).

[3] ISO/IEC IS 21000-3, Digital Item Identification, July 2002 (N4939).

[4] MPEG-21 Requirements Document V.2, December 2003 (N6264).

[5] MPEG-21 Rights Expression Language: Enabling Interoperable Digital Rights Management, Xin Wang, John R. Smith, IEEE Multimedia, October-December 2004, Vol 11, No. 4

[6] ISO/IEC FDIS 21000-7, Digital Item Adaptation, December 2003 (N6168).

[7] U.S. Department Of Defense. Technical Architecture Framework For Information Management (TAFIM) Volumes 1-8, Version 2.0. Reston, VA: DISA Center for Architecture, 1994. http://www-library.itsi.disa.mil/tafim/tafim.html (1996).

[8] ISO/IEC 14496-1 FDAM-3, IPMP extensions on MPEG-4 systems, October 2002 (N5282)

[9] ISO/IEC FDIS 13818-11, IPMP extensions on MPEG-2 systems, March 2003 (N5608)

[10] IPDR Business Solution Requirements - Network Data Management-Usage (NDM-U) v3.5.0.1, IPDR.org, November 2004

A Service-Oriented Framework to Promote Interoperability Among DRM Systems

Fernando Marques Figueira Filho,
João Porto de Albuquerque, and Paulo Lício de Geus

Institute of Computing, University of Campinas, 13083-970 Campinas/SP Brazil

Abstract. Through the past years, several digital rights management (DRM) solutions for controlled dissemination of digital information have been developed using cryptography and other technologies. Within so many different solutions, however, interoperability problems arise, which increase the interest on integrated design and management of these technologies. Pursuing these goals, this paper presents a framework which aims at promoting interoperability among DRM systems, using a service-oriented architecture (SOA) and a high-level policy modeling approach.

1 Introduction

Digital Rights Management is a collection of technologies that enables controlled dissemination of digital information. Today, the majority of DRM applications are used in copyrighted content distribution, such as movies and music, but it is expected that those technologies will also benefit, in a near future, small content producers and individuals who intend to securely distribute their own information.

Although there have been considerable advances in the area, DRM systems still do not interoperate. There are differences over formats and protocols, as well as difficulties in trying to integrate management while simultaneously operating different DRM systems. Thus, content producers are forced to choose one among available platforms, which affects their content distribution covering. Moreover, the lack of operability can be used to stimulate the monopoly over proprietary software and devices by some vendors, which can be harmful for both users and content producers.

Following this motivation, this paper presents a framework which aims at promoting interoperability among DRM platforms. It is based on the fact that in every platform, the lifetime of contents follows basically the same steps: firstly, it is packaged using cryptography, in order to protect it against unauthorized users. Then, at some moment during content distribution, it is licensed to a specific user or device. A license is a file containing the rights and conditions, described in a platform-specific format, which govern contents' usage by that particular user. Our framework centers those rights and conditions in a single policy-based model, which is generic for every DRM platform.

To that effect, a service-oriented architecture (SOA) is defined, which is responsible for managing those policies and using them to generate licenses in

A. Helmy et al. (Eds.): MMNS 2006, LNCS 4267, pp. 124–127, 2006.

different DRM platform formats. Services are implemented using Web Services, allowing for easier compatibility with most computer architectures and programming languages.

The next section presents a brief of the conceptual models in which our approach is based. The system architecture is analyzed in Section 3 and we conclude this paper with some related work and expectations around future work in Section 4.

2   Policy Model

In this paper, policies are based in an object-oriented model which can be divided conceptually into levels of abstraction, as depicted in Fig. 1. The highest level is based on the role-based access control (RBAC) concepts [1] and its extension, the GRBAC [2].

Through the past 10 years, RBAC has been used to simplify permission management, especially when users are hierarchically organized or when it is possible to identify common characteristics among them. Such scenario is found in various DRM business models (e.g. service subscription or purchasing, membership of a club or organization). Instead of associating rights with each user, we apply rights to *subject-roles*, which in turn are associated with users. In this manner, a small policy set is sufficient to manage a large and complex system. Thus, policies in the abstract level are relatively static and their construction is supported by a graphical tool, similar to the one used in other policy-based management applications [3].

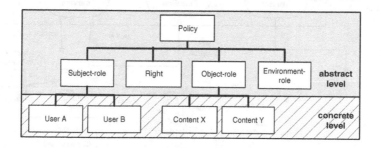

Fig. 1. Policy structure

DRM permissions, however, commonly associate conditions and restrictions to a right (e.g. play, print), based on stateful information. This information is included in the license and used by a particular DRM platform to control, for example, the number of times a user exercises a right, the time interval during which a content can be used, among others. GRBAC extends RBAC through the introduction of *environment-roles*, which are applied to our policy model to incorporate those state-based conditions and restrictions. GRBAC also defines *object-roles*, which are used to group contents and build policies based on their characteristics, such as type (audio, video etc.) and confidentiality level.

The second abstraction level carries concrete entities from a DRM system (e.g. users, contents) and holds a much more dynamic behavior. While the upmost level is updated by human intervention by means of a graphical editor, the second level is updated by framework services according to the external DRM system activity. The architecture that comprehends these services and its functioning are covered in the next section.

3   Framework Architecture

The framework proposed in this work has a service-driven architecture composed by five services. Some are platform-dependent and interface DRM systems with which the framework operates, while others interact with the policy database, as depicted in Fig. 2.

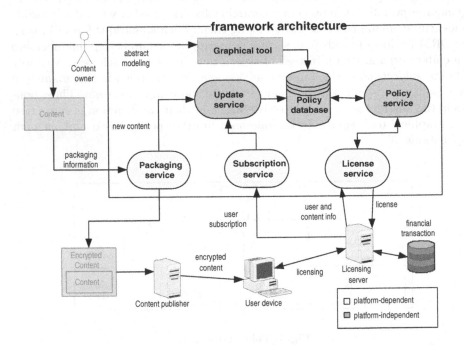

Fig. 2. Framework architecture

In the beginning of the content lifetime it is supposed that abstract policies have already been defined through the graphical tool. After finishing the abstract modeling step, the content can be packaged using a platform-specific *packaging service*, which receives a plain file and references to which *object-roles* that content will be associated with. The *packaging service* then requests the update to the *update service*.

On the other hand, users interact by licensing content on a payment-basis or, when there is no financial transaction involved, by only subscribing to new services and having their access levels changed (e.g. when a company that uses DRM to manage classified documents hires a new employee). In these cases, the *subscribing service* receives user information and references to which *subject-roles* that user will be associated with.

Finally, when a license has to be generated in a specific platform format, the licensing server contacts the *license service* which serves that particular platform, passing user and content identifications, as well as some other platform-specific information. The *license service*, in turn, contacts the *policy service*, which searches the database for all policies related to those user and content, returning the results. The *license service* then interprets the returned policies and generates a license.

4   Related and Future Work

Some recent work analyze interoperability issues, sometimes proposing solutions, as in Sun's project called DReaM [4], which also employs a service-oriented architecture. However, none of them uses a policy-based management approach or any abstract modeling technique.

The proposed architecture aims at providing interoperability through a cen-tered, platform-independent policy model, which interfaces to other systems using specialized services that will be implemented using Web Services. The conceptual division of policies in two layers allows for a system view with an appropriate abstraction level. The high-level policy design is also supported by a graphical editor, to be developed using Java and applying the visualization improvements used in [3].

References

1. Ferraiolo, D., Kuhn, R.: Role-based access control. In: Proceedings of 15th NIST-NCSC National Security Computer Conference, Baltimore, MD (1992)
2. Covington, M.J., Moyer, M.J., Ahamad, M.: Generalized role-based access control for securing future applications. In: 23rd National Information Systems Security Conference Proceedings. (2000)
3. Porto de Albuquerque, J., Isenberg, H., Krumm, H., de Geus, P.L.: Improving the configuration management of large network security systems. In: Ambient Networks: 16th IFIP/IEEE International Workshop on Distributed Systems: Operations and Management, DSOM 2005, Proceedings. Volume 3775 of Lecture Notes in Computer Science., Berlin Heidelberg, Germany, Springer-Verlag (2005) 36–47
4. Fernando, G., Jacobs, T., Swaminathan, V.: Project DReaM - An Ar-chitectural Overview. White Paper. Open Media Commons. Available at: http://www.openmediacommons.org/ (2005)

A Distributed Hierarchical Structure for Object Networks Supporting Human Activity Recognition

Venet Osmani[1], Sasitharan Balasubramaniam[1], and Tao Gu[2]

[1] Telecommunications Software and Systems Group
Waterford Institute of Technology
Ireland
[2] Institute for Infocomm Research
Singapore 119613
{vosmani, sasib}@tssg.org, tgu@i2r.a-star.edu.sg

Abstract. Pervasive environments will witness heterogeneous smart embedded devices (e.g. sensors, actuators) integrated into user's living environment (e.g. smart homes and hospitals) and provide a multitude of information that can transparently support user's lifestyle. One promising application resulting from the management and exploitation of this information is the human activity recognition. In this paper we briefly describe our activity recognition architecture and focus on an important management component of this architecture using the concept of object networks. We explore how object networks can integrate various sensor networks and heterogeneous devices into a coherent network through embedded context and role profile and at the same time support distributed context reasoning. The paper also describes the mechanisms used to eliminate and refine context information that is deemed irrelevant due to user behaviour changes over time, by employing the idea of role fitness.

1 Introduction

The proliferation of wireless devices accompanied by multitude of multimedia applications and services has led to the requirement of responsive environments that can dynamically detect and adjust user activities. One current trend in pervasive systems to achieve dynamic user and application support is the ability to infer user's activities from various devices that supports user's lifestyles. Examples of such devices may include mobile devices (e.g. PDA, mobile phones) or embedded micro device networks (e.g. sensors, actuators). Activity recognition has huge potential in supporting users and pervasive applications by automatically deducing the current activity of a user from various objects in the environment coupled with monitoring user's history of interaction. Although in recent years, tremendous amount of focus has been applied towards evaluating context information to support activity recognition, many of these solutions have relied on centralised context gathering and reasoning techniques. For example in [1], authors rely on centralised architecture with a single level of abstraction of context information to process information regarding

A. Helmy et al. (Eds.): MMNS 2006, LNCS 4267, pp. 128–133, 2006.
© IFIP International Federation for Information Processing 2006

user manipulated objects. Each object is instrumented by a RFID tag that is detected using a glove-mounted RFID reader. Another initiative in activity recognition based on low-level sensors has been recorded in [2], where authors have implemented a largely centralised model that has been deployed in a hospital environment. Guralnik and Haigh [3] follows a similar centralised model for observing human behaviour patterns, where the system only deals with one type of sensor (motion detection sensors) that have been used to gather information in a number of living environments. However, centralised context evaluation has a number of drawbacks which includes (i) performance constraints due to bottleneck effects, and (ii) all context rules and reasoning are centrally located and must cater for devices (e.g. sensors, PDAs, actuators) of various types in the environment.

Our activity recognition system counters these drawbacks by employing a decentralised architecture where reasoning mechanisms for activity recognition are performed collectively from different objects and sensors within the environment through local interactions. The technique for activity inference process is realised by the Activity Inference Engine (AIE) that functions in parallel with the Activity Map (AM) [4]. The AM is a repository of all the activities a user has been engaged in the past and is continually refined to support dynamic user behaviour. The activity inference is performed through an Object Network which provides a platform for processing and aggregating heterogeneous context information that is not only limited to hand held devices that users interact with, but also sensor networks surrounding the user's living environment. Through context information found embedded as profile roles in devices and role to role interaction between heterogeneous devices, a more accurate activity recognition process can be achieved. The rest of this paper is organised as follows: Section 2 describes the Object Network and the interaction process and collective reasoning from various objects within the network to deduce user's activity. Lastly section 3 presents the conclusion.

2 Object Networks

In order to deduce user's activity the AIE typically evaluates information from various objects being manipulated as well as information deduced from these objects describing the state of the environment. The idea of the object networks is a network of heterogeneous devices within an environment that have the ability to interact locally in a peer-peer fashion, while supporting distributed context reasoning. Object networks provide a supporting platform for the generation of the required context information to deduce user's activities using a hierarchical structure as shown in Fig 1. The object networks used to efficiently gather context information from sensors do not only support deduction of activities but also provide support for application adaptability for corresponding activities. An important requirement for context information generated from the object network is that it should be at a *high level of abstraction*. Generally deducing high-level context information is challenging since it requires combination of multiple and heterogeneous context sources on top of a processing logic to deduce such information as described by many authors [5] [6] [7].

The AIE is typically contained within the device, directly supporting the application of the user and is usually selected based on the object (leader object) with the highest capability in the environment (e.g. PDA). The selection is based on an optimisation election process. As shown in Fig. 1, the various objects also house an object profile. The object's profile plays an important role in the discovery, election as well as interaction with peer objects since it describes object's capabilities and *roles* the object can be engaged in.

2.1 Object Role(s)

As we briefly mentioned above, in addition to other properties such as identity and communication capabilities, an object will also assume a specific role. In the most simplistic case a role will specify the type of context information that an object is designed to provide. For example a temperature sensor plays the role of 'providing temperature' since this is the object's primary designation. However objects in the environment that the user interacts with are typically much more complex to be described by a single role. For instance a PDA may fulfil multiple roles, from a simple text editor, up to a web server [8].

The role assigned to an object will remain in effect for the duration of the lifetime of the object, unless object's functionality is changed. Particular objects fulfil their role by utilising information generated within the object, for instance information acquired from onboard sensors, while other objects may in addition require information from external sources. As shown in Fig. 1 we collectively deduce the user's activity from the collection of role interactions between the various objects forming a hierarchical structure.

There are two types of object roles that support the AIE of the leader object which includes the *self-sufficient role* and the *conditional role*. Self-sufficient role objects are typically designed for a specific task and provide only a single layer abstraction of context information (e.g. temperature sensor gives only temperature information).

However, a single level of abstraction of context information is typically insufficient, except in the most simplistic instances. Generally multiple levels of abstraction are desired that involves combining context information from multiple and diverse sources. Thus the objects adhering to self-sufficient role are not particularly apt for this task. An object role that cannot be fulfilled unless the precondition of acquiring external information generated outside the object holds true is known as the conditional role. Certainly both the self-sufficient role and the conditional object role play a major part in increasing the level of abstraction of context information; however the conditional role has a higher influence on this process.

The description of each conditional role within an object semantically defines the type of context information that the object requires in order to fulfil the role in question. In addition, the role description also specifies a set of rules that process the required context information and provide the basis to increase the level of abstraction of this information. For each type of context information required to fulfil a particular role, the object sends a query to the object network semantically describing the type of context information it requires. As soon as the query is propagated through the network, objects providing the requested type of information respond and a '*depends_on*' relationship is established between the requesting object's role and the respondent object's role. Generic representation of this relationship between the role of the requesting object for context information O_a and the role of the context information provider object O_b can be described in terms of Z specification as follows (assuming each object has one role):

$$((O_a, r_a) \wedge (r_a \rightarrow T_r)) \Leftrightarrow \exists (O_b, r_b) \bullet (r_b \rightarrow T_p \wedge T_r \equiv T_p) \wedge (O_a \neq O_b)$$

Therefore a dependency relationship between two objects can be established if and only if exists an object O_b engaged in a role r_b providing information of type T_p which is identical to the type of information T_r requested by a role r_a within the object O_a such that the roles in question are housed in different objects. The last clause of the equation is a 'sanity check' that guards against an object establishing a dependency relationship with itself.

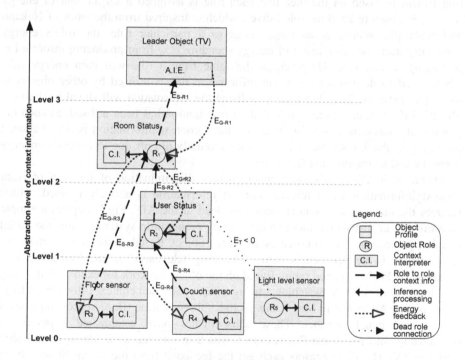

Fig. 1. Objects' role to role interaction

The fact that an object engaged in a conditional role requires information from peer objects that may be engaged in their respective conditional roles creates the necessity to establish a hierarchical context information processing structure. The conditional role will assume at least two levels of abstraction of context information. In instances when the information provider objects specified in the conditional role are further engaged in their respective conditional roles, the context processing and in effect the abstraction level of context information is further augmented with multiple levels of abstraction of context information. In order to realise the hierarchical context processing structure we establish sub-networks within the overall object network. The sub-networks are smaller entities with a structure similar to the overall object network, containing a *central* object (similar to the leader object) that requests information from the *edge* objects forming the sub-network. As a consequence the object network shifts away from a two level architecture, creating multiple tiers whereby each tier has the effect of increasing the level of abstraction of context information.

2.2 Role Fitness for Activity Recognition Refinement

User behaviour is considered dynamic and results in activity characteristics that continually change to support user experience. In order to refine the activity inference process to suit the changing user behaviour we employ the *role fitness* concept. The role fitness is based on the idea that each role is assigned a set amount of energy $(E_T = E_G - E_S)$ used to keep the role 'alive', which is inspired from the work of Nakano and Suda [9]. Whenever an object engages a particular role, the role's energy constantly decreases as a result of energy spending (E_S) from producing information pertaining to that role. However, at the same time a role will gain energy (E_G) whenever the role generates relevant information that is utilised by other objects in the object network. The object's role utilising the information will signal back to the object's information generator role that its information has been utilised, as such the amount of energy increases. In the event that a context information is not evaluated based on feedback response, the role will eventually die off due to excessive energy spending and no energy gain $(E_T <= 0)$.

Therefore in effect this mechanism enables discrimination of the non relevant context information while relevant context information is actively utilised, which reduces the unnecessary context evaluation load at the AIE. Fig.1. depicts a simple scenario of our distributed hierarchical activity recognition systems. After the initial election process the TV is chosen as the leader object. Based on the activity map, the next activity to be deduced is either "watching television" in the living room or "chatting on PC" in the bed room. Based on the object network for each activity, only the object network from "watching a tv" receives events and context information from peer objects. As shown in Fig. 1, this includes three objects (floor sensor, light level sensor, and couch sensor). These sensor roles have established *depends on* relationship with the Room Status object, which in turn has *depends on* relationship with the TV. The three sensors each get the feedback from the Room Status object increasing the energy levels, while the Room Status object gets feedback from the leader object (TV). However user habits have changed and now the user prefers to watch the sports programme with the lights off. Such event is detected by the room status object, resulting in the information about the light levels becoming irrelevant $(E_T <= 0)$. The light sensor's ratio between the gained energy and spent energy is increasing in favour of the latter, causing the light level sensor role to slowly deplete and eventually die, as shown in Fig. 1.

3 Conclusion

In this paper we have demonstrated mechanisms to manage objects within an environment in order to deduce the activity of a user. We have combined several ideas to achieve the required functionality which includes embedding roles into various devices and supporting role to role interaction to allow high-level context abstraction and distributed context reasoning. While object roles provide the necessary information to enable the activity inference process, the concept of role fitness augments this process to make it dynamic and adjustable to changing user behaviour by effectively eliminating roles that are deemed irrelevant to deduce user activity. We are in the process of implementing the architecture and deploying this into a simulated environment to evaluate the overall design for our current work on activity recognition supporting health-care [4].

References

1. Matthai Philipose, Kenneth P.Fishkin, Mike Perkowitz, Donald J. Patterson, Dieter Fox, Henry Kautz, and Dirk Hähnel, "Inferring Activities from Interactions with Objects", IEEE Pervasive Computing, vol. 03, no. 4, pp. 50-57, October-December, 2004
2. Jakob Bardram and Henrik Bærbak Christensen, "Open Issues in Activity-Based and Task-Level Computing", First International Workshop on Computer Support for Human Tasks and Activities, CfPC PB-2004-60, 2004
3. Valerie Guralnik and Karen Zita Haigh, "Learning Models of Human Behaviour with Sequential Patterns", Proceedings of the AAAI-02 workshop 'Automation as Caregiver', pp. 24-30, 2002
4. Venet Osmani and Sasitharan Balasubramaniam, "Context Management Support for Activity Recognition in Health-Care", Pervasive 2006 Workshop Proceedings (ISBN 978-3-00-018411-6), pp.453-465 T. Strang, V. Cahill, & A. Quigley, eds in 3rd International Workshop on Tangible Space Intiative, 2006
5. Anind K. Dey, "Providing Architectural Support for Building Context-Aware Applications", PhD Thesis, Georgia Institute of Technology, 2000
6. Harry Chen, Tim Finin, and Anupam Joshi, "An Intelligent Broker for Context-Aware Systems", Adjunct Proceedings of Ubicomp 2003, Seattle, Washington, USA, 2003
7. Tao Gu, Xiao Hang Wang, Hung Keng Pung, and Da Qing Zhang, "An Ontology-based Context Model in Intelligent Environments", In Proceedings of Communication Networks and Distributed Systems Modeling and Simulation Conference, pp. 270-275, 2004
8. Newmad Technologies, "Newmad Technologies AB", Accessed 04/08/2006, Available from http://www.newmad.se/picowebserver.html, 2006
9. T. Nakano and T. Suda, "Self-organizing network services with evolutionary adaptation", IEEE Trans Neural Netw. 2005 Sep;16(5):1269-78, 2005

A Bandwidth-Broker Based Inter-domain SLA Negotiation

Haci A. Mantar[1], Ibrahim T. Okumus[2], Junseok Hwang[3], and Steve Chapin[4]

[1] Department of Computer Engineering, Gebze Institute of Technology, Turkey
[2] Department of Computer Electrical Engineering and Computer Science, Syracuse University, Syracuse, NY, USA
[3] Seoul National University, Seoul, Korea
[4] Department of Electronics and Computer Education, Mugla University, Turkey

Abstract. This work presents an Inter-Domain Bandwidth Broker (BB) based Service Level Agreements (SLAs) Negotiation Model for Differentiated Services (DiffServ) networks. A BB in each DiffServ domain handles SLAs on behalf of its domain by communicating with its neighboring peers. The proposed model uses a destination-based SLAs aggregation to increase signaling and state scalability, and it uses a BB-based inter-domain routing to increase resource utilization. The experimental results are provided to verify the achievements.

1 Introduction

The Differentiated Services (DiffServ) has become a key technology in achieving Quality of Services (QoS) in the Internet. DiffServ aggregates individual flows into different traffic classes at the edge of the network, and core routers within the domain forward each packet to its next hop according to the per-hop behavior (PHB) associated with the traffic class of the packet. Diffserv requires no per-flow admission control or signaling and, consequently, routers do not maintain any per-flow state and operation, and this greatly improves the scalability. However, due to the lack of admission control and signaling, DiffServ does not provide QoS guarantees to individual flows. Recently, many studies have focused on providing QoS guarantees in a domain. However, delivery of end-to-end QoS to support end-user applications requires the resource reservation in all the domains along the path. The resource reservation among different administrative domains is handled with service level agreements (SLAs) negotiation.

The Bandwidth Broker (BB) model [2] is a strong candidate for SLAs negotiation among DiffServ domains. As a central logical entity in each DiffServ domain, a BB is mainly responsible for the inter-domain SLA negotiation for its entire domain. The BB makes policy access, resource reservation and admission control decisions on behalf of its entire domain.

In [1], we present a Simple Inter-BB Signaling (SIBBS) protocol. A BB uses SIBBS to communicate with its peers to reserve resources for its inter-domain QoS traffic. SIBBS employs a core tunneling model, in which a pipe is pre-established between each possible source and destination domain and carries the

A. Helmy et al. (Eds.): MMNS 2006, LNCS 4267, pp. 134–140, 2006.

traffic of a particular class. Pipes are identified by a destination domain IP prefix and by DiffServ Code Point (DSCP). After a pipe is established, all individual reservations for particular destination and DiffServ class are multiplexed into the same pipe. The states of border routers and signaling messages exchanged between BBs are pipe-based. All the reservations in a pipe are considered as a single reservation from the intermediate BBs' point of view.

By aggregating individual reservations into an existing pipe, SIBBS significantly reduces the inter-BB signaling load and admission control time compared to per-flow/per-reservation schemes. Although SIBBS can be effectively used for small-scale networks such as VPN applications, it has a serious scalability problem when it is applied to large-scale networks such as the entire Internet. The number of pipes in core transit domains scale with $O(n^2)$, where n is the number of domains or networks in the Internet. When it is applied to the entire Internet, this number might be too high to be managed. Another problem is that the numbers of inter-BB signaling messages and states maintained by a BB are proportional to the number of pipes. This causes processing, signaling and state scalability problems.

In this work we enhance SIBBS by adding a destination-based aggregation to make the inter-domain SLAs negotiation scalable and efficient. As the only entity responsible for QoS resource control in a domain, a BB uses this protocol to reserve/negotiate QoS resources with its peers located in the neighboring domains for its inter-domain QoS traffic.

2 Destination-Based Inter-domain SLAs Negotiation

The idea of destination-based aggregation is very simple. A provider domain's BB aggregates its customers' (upstream domains) pipe requests of the same destination region and QoS class, and makes a single reservation (setup a single pipe) with its downstream domain on behalf of all of its customers. We name this scheme enhanced SIBBS (eSIBBS).

Consider Figure 1, where $S1, S2, S3$ represent source domains, $T1, T2$ represent transit domains and D represents a destination domain. For simplicity, we assume that there are no end hosts located in transit domain and that requests are directed from Si to D. We also assume that requests are based on bandwidth demands. As shown, source domains $(S1, S2, S3)$ establish a pipe to transit domain $(T1)$ for the traffic destined for D. The BB of T1 aggregates its customers (S1, S2, S3) requests and establishes a pipe to D. The resources of the pipe between $T1$ and D is shared by all the customers of $T1$, and the resource negotiation is based on the pipe utilization. To reduce the numbers of signaling messages with D, the $T1$ can make the reservation size more than the current demand (the sum of the demand from $S1, S2,$ and $S3$). The BB of $T1$ can grant its customers requests as long as there is available resources in the pipe with D.

In the above aggregation scheme, we assume that the aggregated traffic follows a single path, determined by BGP-4, towards a destination (as if there is only a single provider for each destination). The resource provisioning and reservation is

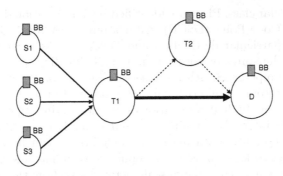

Fig. 1. A simple multi-provider network

performed based only on the resource availability on that path. While this single path based (BGP-4 based) aggregation scheme increases scalability, it may not be sufficient for resource utilization efficiency.

Assume that only the path $T1 - D$ has limited resources (Figure 1). When the traffic rate of the path $T1 - D$ reaches its maximum capacity, the reservation requests coming from source domains will be rejected even if the alternative path, $T1 - T2 - D$, has available resources. This is because the traffic from $T1$ to D is routed through the link $T1 - D$, which is the path given by BGP-4.

To alleviate this problem, we enhance the BB to play a role in inter-domain path selection. Among the possible candidate paths, provided by BGP-4, the BB selects a path that has available resources by signaling the candidate BBs. For example, $T1$ has two alternative paths ($T1 - T2 - D$ and $T1 - D$) to send its traffic to D. When the BB of $T1$ needs resource reservation to D, it can send a request to the BBs of D and $T2$ asking for resources. Upon receiving the request, the BBs D and $T2$ of send reply back to the BB of $T1$. The BB of $T1$ then chooses the path that has the requested resources. In case both of them have available resources, it can choose the least costly one.

Selecting a path based on the resource availability can significantly increase the resource utilization. But, sometimes it may not be possible to send the whole aggregated traffic through a single path. In this case, a domain can send the aggregated traffic over multiple paths. For example, $T1$ can split and send its traffic over the paths $T1 - D$ and $T1 - T2 - D$.

To split the aggregated traffic over multiple paths in a way that avoids out-of-order packet delivery, we use a hashing based scheme. The hashing can be done based on source and destination IP addresses, and possibly other fields of IP header. In hashing model, the traffic will be first distributed into N bins by using module-N operation on the hash space. If the total traffic rate is X bps, each bin approximately receives the amount of X/N bps. The next step is to map bins to paths. The number of bins assigned to a path is determined based on its load portion.

With the hashing-based scheme, the number of states that border routers, which performs splitting, maintain is independent of the number of flows constituting the aggregation. By receiving the load portion of each path from the BB, the routers just need to perform forwarding based on the result of module-N operation.

3   Evaluation Results

In this section, we present simulation results to verify that the proposed model is robust, scalable and efficient. Figure 2 illustrates the configuration of the test-bed, which consists of 100 source domains $(S1, S2, ..., S100)$, 10 destination domains $(D1, D2, ..., D10)$ and five transit domains $(T1, T2, T3, T4, T5)$. Each source domain and destination domain have 10 end hosts connected. Resource reservation requests are from source domains to the destination domains.

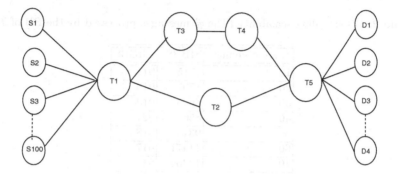

Fig. 2. BB Test-bed

We used real-time traffic traces collected from real networks for our experiments. We used traffic traces provided by CAIDA [7], which has advanced tools to collect and analyze data based on source and destination Autonomous Systems (AS) and traffic types for short time intervals. Source domains sent the traffic to the destination domain based on the traced data characteristics. In order to imitate the traced traffic characteristics, hosts located in source domains were configured to send UDP traffic with a variable average rate that reflects the traced data. The experiments are performed for virtual leased line (VLL) services [1][4]. The traffic rate of an individual request and a pipe is determined with the parameter-based scheme where each host requests reservation with the peak rate. All the experiments were run for 25-minute periods. The duration of a reservation was exponentially distributed with a mean of 1 minute. The reservation rates varied over the duration of the experiment, based on the rate profile originated from the traced data.

Table 1 shows the number of state messages that a BB (the BB of $T1$ in this experiment) maintains in case of SIBBS and eSIBBS. The table shows that a BB can be a potential bottleneck in case of SIBBS when the number of source-destination pairs are large. On the other hand, this number is proportional to the number of destination domains in eSIBBS. The BB signaling scalability

Table 1. BB state scalability

# src.domains	SIBBS	eSIBBS
10	865	493
20	1877	497
30	2911	505
40	3818	507
50	4873	508
60	5756	509
70	6871	509
80	7953	509
90	8909	509
100	9929	509

Table 2. BB signaling scalability (The # messages processed by the BB of $T1$)

# src domains	SIBBS	eSIBBS
10	1715	911
20	3707	915
30	5915	913
40	8567	921
50	9312	918
60	11471	917
70	13366	913
80	15213	915
90	17643	911
100	19789	909

Table 3. Border router state scalability

# src. domains	SIBBS	eSIBBS
10	187	93
20	191	97
30	285	105
40	391	108
50	487	109
60	579	109
70	678	109
80	785	109
90	892	109
100	991	109

results (the number of messages processed by a BB) are given in Table 2. Table 3 illustrates the comparison of SIBBS and eSIBBS in terms of border router state scalability. As depicted, when the number of sessions in each stub domain increases, the number of states in border routers increases. In case of eSIBBS,

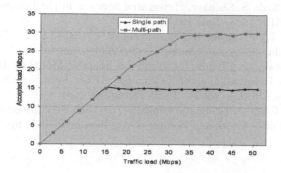

Fig. 3. Accepted load in single and multi-path scheme

the number of states remain almost constant. (This measurement was done in the egress router of $T1$.)

In the last experiment (Figure 3), we evaluated the inter-domain multi-path scheme. We configured the paths $T1 - T3 - T4 - T5 - Di$ and $T1 - T2 - T5 - Di$ with 15Mbps and the rest of the links in the network with 30Mbps capacity. In single path case (when BGP-4 was used), the BB of $T1$ sends all its traffic via $T2$. When that path reaches its capacity, the BB of $T1$ rejects all the incoming requests. In multi-path case, the BB of $T1$ accepts the requests up to 30Mbps. Because it sends through both providers ($T2$ and $T3$).

4   Conclusion and Future Work

This work shows the significant role a BB can play for service providers, and addresses many outstanding questions about the BB model. The experimental results show that an ISP can substantially improve its resource utilization, thereby increase its revenue, while requiring minimal changes in the underlying infrastructure. The scalability and utilization results provide the basic guidelines that an ISP should consider in defining SLAs.

In this work we have relied on VLL services (EF), however, it is obvious that there is a potential need for other services, so we will enhance SIBBS protocol to support other DiffServ classes (e.g., Assured Forwarding). In the future, we will also investigate the marketing aspect of SLAs among ISPs.

References

1. QBone Signaling Design Team, "Simple Inter-domain Bandwidth Broker Signaling (SIBBS)", http://qbone.internet2.edu/bb/ work in progress.
2. K. Nichols, V. Jacobson, and L. Zhang. " A Two-bit Differentiated Services Architecture for the Internet" RFC 2638, July 1999.
3. S. Black et al., "An Architecture for Differentiated Services," RFC2475, Dec. 1998.
4. V. Jacobson, K. Nichols, K. Poduri, "An Expedited Forwarding PHB," RFC2598, June 1999. A

5. R. Braden, D. Clark, S. Shenker, "Integrated Services in the Internet Architecture: an Overview," RFC 1633, June 1994.
6. Y. Rekhter, T. Li, "A Border Gateway Protocol 4 ", RFC 1771, March 1995.
7. http://www.caida.org/dynamic/analysis/workload/sdnap/.
8. H. Mantar, J.Hwang, S. Chapin, I. Okumus "A Scalable Model for Inter-Bandwidth Broker Resource Reservation and Provisioning", IEEE Journal on Selected Areas in Communications (JSAC), Vol.22, No.10, December 2004
9. D. Di Sorte, G. Reali: "Pricing and Brokering Services over Interconnected IP Networks", Journal of Network and Computer Applications, 2005, pp. 249-283.

Efficiency of Loss Differentiation Algorithms in 802.11 Wireless Networks

Stephane Lohier[1] and Yacine Ghamri Doudane[2]

[1] LIP6 - University of Paris VI, 8, rue du Capitaine Scott, 75015 Paris – France
[2] IIE-CNAM, 18, allée Jean Rostand, 91025 Evry Cedex – France
stephane.lohier@lip6.fr, ghamri@iie.cnam.fr

Abstract. Loss Differentiation Algorithms (LDA) are currently used to determine the cause of packet losses with an aim of improving TCP performance over wireless networks. In this work, we are interested in the distinction between losses due to interferences on 802.11 links from those due to congestions. To this end, we compare different LDA schemes existing in the literature and a proposal of a cross-layer LDA which use MAC parameters. The efficiency of our solution is then demonstrated through simulations.

1 Introduction

Due to frequent signal failures, the performance of TCP in the 802.11 networks can be significantly degraded, particularly in SOHO (Small Office Home Office) environments or in public points of distribution (Hot Spot) with a wireless last link. Indeed, TCP is conceived for wired networks and is not adapted to react to losses on wireless links. Loss Differentiation Algorithm (LDA) are thus necessary to determine if the segment loss is due to short signal failures caused by interferences in the vicinity of the wireless station and its AP (Access Point) or by congestions due to heavy traffic. In this work, we compare, in the selected context, different LDA schemes and a proposal of an alternative LDA based on a MAC layer parameter which is specifically adapted to the 802.11 links. A simulation set makes it possible to evaluate the effectiveness of the various solutions.

2 The Different LDA Schemes

As proposed in the literature [1-3], the differentiation decision can be obtained based on TCP parameters, namely packets inter-arrival times, *cwnd* (congestion window) and *RTT* (Round Trip Time). The LDA using packet inter-arrival times at the receiver [1] are unsuited here as they consider that the resolution is always completed at the TCP layer and thus do not take into account the MAC layer loss-recovery present in the 802.11 standard. A set of preliminary measurements (Fig. 1) confirms that, with a first loss-recovery level at MAC-layer, the inter-arrival times are not significant to differentiate theses two type of loss. Even for important interference or congestion rates (greater than 70%), at the saturation limit of the wireless link, the obtained values remain relatively close.

A. Helmy et al. (Eds.): MMNS 2006, LNCS 4267, pp. 141–144, 2006.

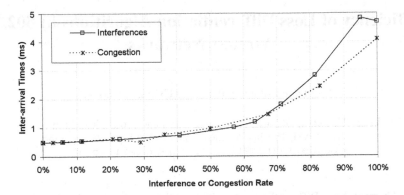

Fig. 1. Efficiency of the Inter-arrival Time LDA on an 802.11 link

For losses caused by relatively important interferences and unsolved by MAC retransmissions, TCP reacts as for congestions: *cwnd* reduced following the reception of three duplicate ACKs. As the *cwnd* variable is updated each time a segment-loss occurs, it became highly variable. The simulation study realized in [2] showed that, due to its variability, *cwnd* is not an accurate and selective variable for differentiation.

Finally, in a context with a wireless last hop, *RTT* is the variable presenting the most significant variations in the event of short signal losses compared to congestions. We thus selected three TCP-layer LDA schemes based on comparisons of current *RTT* values with different thresholds and on filters giving more or less weight to the recent samples: the *Vegas Predictor* scheme [2], the *Spike* scheme [1] and the *Flip Flop Filter* [3]. Then, rather than using only TCP-layer parameters which do not take into account the 802.11 specificities, we suggest to use a cross-layer approach as an alternative to conventional TCP-Layer LDA schemes.

The idea of our alternative algorithm is to count the number of MAC retransmissions for each of the *n* segments composing the current TCP window when the TCP layer is alerted by the reception of three duplicated ACKs. As described in Table 1, if for one of these segments at least, the number of MAC retransmissions (*RetryCount*) is equal to the *Retry Limit* threshold (its default value is 6 for all the 802.11 equipments), we consider that the loss is due to interferences and not to TCP congestion. Indeed, in the case of congestion, the surplus of segments is eliminated from the queue of the concerned node and MAC retransmissions are theoretically not used; inversely, in case of persistent interferences, the segment is dropped by the MAC layer after reaching the *Retry Limit* threshold. This algorithm assumes that for all the not acknowledged TCP segments, the value of *RetryCount* is stored. The *ACKFailureCount* counter available in the 802.11 *Management Information Base* (MIB) gives the number of times that an expected ACK is not received and consequently the value of *RetryCount*.

Note that while the TCP sender is not a wireless host and that the TCP flow is forwarded to the wireless receiver through an AP, an additional stage is necessary. The *LDA_Estimator* is first set at the AP's MAC layer. Then this latter informs the TCP sender by setting consequently the *ELN* (*Explicit Loss Notification*) bit of the TCP header in the ACK segments (i.e. *ELN=LDA_Estimator=1* in case of interferences). The loss differentiation is finally made at the TCP sender when receiving three duplicated ACKs.

Table 1. Cross-layer LDA

if (3 dup ack) **then**	// loss indication in TCP algorithm
LDA_Estimator = 0	// initial value for congestion
for (i = 0 ; i ≤ n ; i ++)	// for all the not acknowledged segments
if (RetryCount = RetryLimit) **then**	// segment is dropped, probably a short loss
LDA_Estimator = 1	// set value for interferences
end if	
end for	
end if.	

3   Simulation Results

In order to realize a comparative study among the 4 LDA schemes, a set of simulations with a wireless last link undergoing congestions or interferences have been realized. The TCP flow is thus established between a server and a wireless station through its AP. Interferences are caused by the transmission on the same channel of a CBR/UDP flow between two other wireless stations out of the AP coverage and interferences areas. As we deactivated the RTS/CTS mechanism for both transmissions, the AP will not detect CBR transmissions and will thus transmit its TCP segments toward the receiver which is located in the interference area. The duration and the frequency of the interferences will vary according to the size of the packets and the rate of the CBR source (with 1000Bytes packets sent at 1800packets/s the wireless link is completely saturated, which correspond to 100% on the curves). Let us note that the simulated interferences and so the packets losses are carried out in a scenario close to reality (short losses are often caused by transmissions in the same frequency band) and not with a theoretical packet error rate as inaccurately used in most studies. Another CBR/UDP flow is established between the server and a fourth wireless station in order to saturate the AP and induce congestions (for this flow, a rate of 3500packets/s with 1000Bytes per packet is necessary to saturate the link).

The simulation results presented in Fig. 2 and Fig. 3 show the accuracy (the percentage of correctly classified losses) of the four LDA schemes according to the interference or congestion rates. For the *Vegas predictor* scheme, we observe that the losses due to low interference rates or high congestion rates are badly classified (less than 60%). This is especially due to the evolution of *cwnd*, which is in these cases inadequately used in conjunction with *RTT* to compute the *Vegas predictor*. The *Spike* scheme, only based on *RTT* variations, gives slightly better results: accuracy higher than 80% in the majority of the cases. The badly classified losses are more random and are mainly due to the calculation mode of the Spike's thresholds. The *Flip Flop* filter is not very efficient, particularly for losses due to interferences. The used algorithm employs many parameters difficult to regulate. Finally, the proposed cross-layer LDA scheme gives the best results. For congestions, there are almost no MAC retransmissions and the *Retry Limit* threshold is never reached, which gives 100% of correctly classified losses. For interferences, some losses are badly classified when the segment is finally received with the last attempt. However accuracy remains in all the cases higher than 90%.

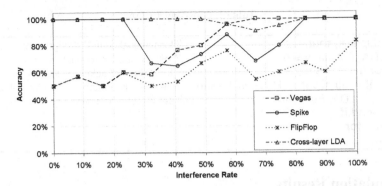

Fig. 2. Accuracy of the 4 LDA schemes with Interferences

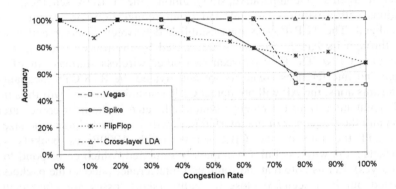

Fig. 3. Accuracy of the 4 LDA schemes with Congestions

4 Conclusion

In this paper, we demonstrated, through comparative simulations, the efficiency of our cross-layer LDA to distinguish congestions from short losses due to interferences in an 802.11 network. This differentiation is a first step to a complete solution integrating an improvement of the TCP behavior. Indeed, in case of interference, the TCP loss recovery mechanism does not have to be triggered which allows the source to achieve higher transmission rates when the MAC channel is rapidly restored.

References

1. S. Cen, P. C. Cosman, and G. M. Voelker, "End-to-End differentiation of congestion and wireless losses," IEEE/ACM Transactions on Networking (TON), Volume 11, Issue 5, October 2003.
2. S. Biaz and N. H. Vaidya, "Distinguishing Congestion Losses from Wireless Transmission Losses: a Negative Result," Proc. of IEEE 7th International Conference on Computer Communications and Networks, IC3N'98, October 1998.
3. D. Barman and I. Matta, "Effectiveness of Loss Labeling in Improving TCP Performance in Wired/Wireless Networks," Boston University Technical Report, 2002.

Towards Autonomic Handover Decision Management in 4G Networks*

Joon-Myung Kang[1], Hong-Taek Ju[2], and James Won-Ki Hong[1]

[1] Dept. of Computer Science and Engineering, POSTECH, Korea
{eliot, jwkhong}@postech.ac.kr
[2] Dept. of Computer Engineering, Keimyung University, Korea
juht@kmu.ac.kr

Abstract. Mobility management has become an important issue in 4G networks due to the integration of multiple network access technologies. Traditionally, only the received signal strength has been considered for the vertical handover. However, more considerations will be necessary to satisfy the end user's preferences. In this paper, we propose an Autonomic Handover Manager (AHM) based on the autonomic computing concept to decide the best network interface to handover in 4G networks. AHM decides the appropriate policy for the specific service or application without the user's intervention using the context information from the mobile terminal, the network and the user. We present the context information and the context evaluation function to decide handover based on the user preferences. We then describe the scenario to validate its feasibility using multimedia conferencing service on the mobile terminal.

Keywords: Autonomic Management, 4G Networks, Context Management, Vertical Handover, Mobility Management.

1 Introduction

Recently, the network trend has been moving towards the architectures that support wireless technologies, mobile users, multiple radio access technologies, heterogeneous networks and network convergence. In 4G networks, the users are in need of services using the terminal that supports multi-mode and multi-access, which means that the terminal can use any network regardless of network access technologies (e.g., CDMA, WLAN, Bluetooth and WiBro) or network providers. As a result, an interesting problem has surfaced on deciding the "best" network interface to use at any given moment. It is apparent that the decision should be based on various considerations such as the capacity of each network link, usage charge of each network connection, power consumption of each network interface, battery status of the mobile device and user preferences. They are called as context information from the network, the system and the user profile.

---

* This research was supported by the MIC (Ministry of Information and Communication), Korea, under the ITRC (Information Technology Research Center) support program supervised by the IITA (Institute of Information Technology Assessment)" (IITA-2005-C1090-0501-0018) and by the Electrical and Computer Engineering Division at POSTECH under the BK21 program of the Ministry of Education, Korea.

A. Helmy et al. (Eds.): MMNS 2006, LNCS 4267, pp. 145–157, 2006.

In this research area, mobility management which supports an efficient vertical handover [1], and Always Best Connected (ABC) [2] which provides the seamless service to the user are important aspects in the heterogeneous network environment. Horizontal handover is the process of maintaining the mobile user's active connections as it moves within the wireless network. Vertical handover, or inter-system handover, deals with the handover between different types of networks. Traditionally, research on the handover has been based on an evaluation of the received signal strength (RSS) at the mobile node. However, traditional RSS comparisons are not sufficient enough to make a vertical handover decision, as they do not consider the various available options for the mobile user. Other factors, such as monetary cost, network conditions, mobile node conditions, user preferences and so on, must be considered, as well as the capabilities of the various networks in the vicinity of the user. Thus, a more sophisticated, adaptive and intelligent approach is needed to implement the vertical handover mechanisms to produce a satisfactory result for both the user and the network. In addition, services can be composed of applications that could use different network interfaces. A single connection session for the specific service will be maintained efficiently by the service control. Therefore, a thorough evaluation of the types of services and applications are necessary in making a proper decision on the network interface.

In this paper, we propose the Autonomic Handover Manager (AHM) based on the autonomic computing concept [3, 4] to determine the best network interface to handover in 4G networks. This manager generates a policy to determine the best network interface for the specific service by using the context from the terminal, the network and the user. Then, it maintains by itself to provide seamless service to the user with the self-management mechanism.

The remainder of this paper is organized as follows. Section 2 describes the related work on the handover decision models, context management, and the autonomic computing concept. Section 3 explains our solution approach. Section 4 presents the architecture of AHM and the context evaluation function. Section 5 presents an interesting scenario using our proposed model and evaluates it qualitatively as the user-centric aspect. Finally, conclusions are drawn and future work is discussed in Section 6.

2 Related Work

In this section, we present the previous work and the problems related to the handover decision model. We also describe the context management in mobile environments for the vertical handover and the concept of autonomic computing.

2.1 Handover Decision Models

Related work on the handover decision model has been presented in recent research literature. A policy-based handover scheme has been proposed in [5], where the authors designed a cost function to decide the "best" moment and interface for the vertical handover. However, the cost function presented in that paper is very preliminary and cannot handle sophisticated configurations. The logarithmic function used in the cost function also has difficulty in representing the cost value while the value of the constraint factor is zero (e.g., the connection is free of charge). Another scheme proposed in [6] models the handover with HTTP traffic, but it may have problems with other types of traffic, such as video and audio streaming, where the bandwidth demand is much higher than HTTP traffic. A smart decision model in [7] is proposed to perform vertical handover to the "best" network interface at the "best" moment and is tested on the Universal Seamless Handover Architecture (USHA) [8]. The smart decision model is based on the properties of available network interfaces (link capacity, power consumption and link cost), system information (remaining battery) and user preferences. Although the model presented a detailed example on the USHA test-bed, it was not enough to describe how to achieve the properties and the meaning of cost value. The authors of [9] aim to understand how to define a metric in order to devise a solution that balances the overall cost of the vertical handovers with the actual benefits they bring to the user's networking needs. This way, each mobile user could autonomously apply the handover decision policy, which is more convenient to the user's specific needs. A vertical handover decision function proposed in [10] allows the user to strategically prioritize the different network characteristics such as network performance, user preference and monetary cost. This function is simple and can be easily applied to any vertical handover approach. The authors presented some characteristics for decision function such as the cost of service, security, power requirements, proactive handover, quality of service and velocity. However, this study was not enough to describe how to define each characteristic and to present the example to validate the decision function. Also, it did not consider the cost function in terms of the service or the application.

2.2 Context Management

The context management framework proposed in [11] has optimized the services in the mobile environment based on the context information. This ensured that the correct context information is available at the right place at the right time, and that it handles diverse, dynamic and distributed context information. This was implemented in the prototype [12], where different context formats and exchange algorithms were evaluated in detail. The context information can be applied to the decision model for vertical handover. In general, context information can be static or dynamic and can come from different network locations, protocol layers and device entities.

Table 1. Context Information Classification

	Mobile Devices	Network
Static	User settings and profile	User profile and history
	Application setting	Network location, capabilities and services
	Willingness to pay	Charging models
Static in a cell	Reachable access points	Potential next access point
Dynamic	Type of application	Location information and location prediction
	Application requirements	Network status such as signal strength
	Device status (battery, interface status)	Network traffic load

Table 1 [12] provides a classification of context information to be considered in the mobile environments. This table is clearly just a snapshot, as other types of context information may be considered as well. They presented their context management framework which can collect and process the relevant context information using context repositories, context server, mobile context client and network traffic monitor. They compiled the context information into a decision matrix exploiting context information for the horizontal handover. We have presented the mechanism to apply the context information for the vertical handover more complicated than the horizontal handover.

2.3 Autonomic Computing

Autonomic computing was proposed as a systematic approach to achieving computer-based systems managing themselves without human interventions [3]. An autonomic computing system has four basic characteristics for self-management. They are self-configuring, self-healing, self-optimizing and self-protecting: Self-configuration frees people to adjust properties of the system according to changes of the system and environment; self-healing frees people to discover and recover or prevent system failures; self-optimization frees people to achieve best-of-the-breed utilization of resources; self-protection frees people to secure the system.

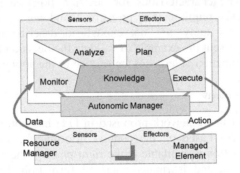

Fig. 1. The Architecture of Autonomic Manager

An autonomic manager is an implementation that automates some management function and externalizes this function according to the behavior defined by management interfaces. As shown in Fig 1, the architecture dissects the loop into four parts that share knowledge. These four parts work together to provide the control loop functionality.

- Monitor function provides the mechanisms that collect, aggregate and filter and report details collected from a managed resource.
- Analyze function provides the mechanisms that correlate and model complex situation. These mechanisms allow the autonomic manager to learn about the IT environment and help predict future situations.
- Plan function provides the mechanisms that construct the actions needed to achieve goals and objectives. The planning mechanism uses policy information to guide its work
- Execute function provides the mechanisms that control the execution of a plan with considerations for dynamic updates.

These four parts communicate and collaborate with one another and exchange appropriate knowledge and data to achieve autonomic management.

3 Solution Approach

We propose a handover decision model using the context information to overcome the limitations of the previously proposed decision models in Section 2. In the mobility management, the decision function to decide a handover is one of the most important functions. It should use the context information of the network, the mobile terminal and the user.

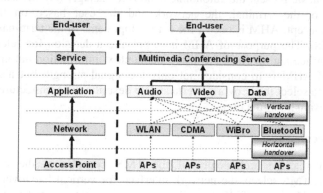

Fig. 2. Layering for general service and multimedia conferencing service

Fig 2 shows the general service scenario on the left and multimedia conferencing service on the right considering the application, network and access point in a heterogeneous mobile network environment. For instance, the end-user is using the multimedia conferencing service on the mobile terminal which can access multiple radio networks using multiple network interfaces. It is composed of multiple applications such as audio, video and data application. Each application succeeds in making a session with one of multiple access networks such as WLAN, CDMA, WiBro and Bluetooth. Then, the mobile terminal connects with one of the access points. There are two kinds of handover; the vertical handover which occurs when a different type of access network is selected, and the horizontal handover which occurs when another access point is selected within the same type of access network. As mentioned earlier, traditionally, the selections of access network and access point are based on the signal strength. In future, more parameters related to QoS, battery and service pricing should be considered because the end-users might use the service according to their needs which could prevent them from being interested in selections of access network and access point.

In order to solve such problems, we propose an autonomic handover manager to make a decision to handover with the context information. The autonomic handover manager is one of the key components which compose the terminal management system. It can monitor, analyze, plan and execute using sensors and effectors. It needs the context evaluation function to choose a policy, using the context information. The context evaluation is represented by the context evaluation matrix accompanied by additional rules and policies. The context information is provided by the context management framework which supports various context repositories distributed in the networks. The context server collects the relevant context information from the context repositories and prepares the evaluation matrix. The mobile terminal evaluates the evaluation matrix for context processing and makes a decision on handover. In the next section, we describe our solution by the approach mentioned above.

4   Autonomic Handover Manager

In this section, we present the autonomic handover manager (AHM) using the context information of the terminal, the network and the user based on the autonomic computing concept. AHM is one of the core components in the terminal management system (TMS). The purpose of AHM is to provide a decision for selecting the best network interface and the best AP using the context information autonomously. The main idea of AHM is to consider not only the signal strength but also the context information to select the network interface. We describe the architecture of AHM and the context evaluation function.

4.1   Architecture

Fig 3 illustrates the architecture of the TMS considering the context server. Traditionally, TMS selects the network interface for vertical handover by the signal strength only. However, TMS for 4G networks requires an innovative architecture that is capable of dynamically selecting the appropriate network through which services can be obtained efficiently in terms of cost and QoS in a transparent manner. Our proposed architecture which has the AHM is able to use the context information from the mobile node and the context server. The context server collects the network context information from the respective operations and support systems (OSS). Currently, this is not realistic because it is difficult and rather impossible to share the information among the service providers. We assume that this will be possible in a 4G network and service environment.

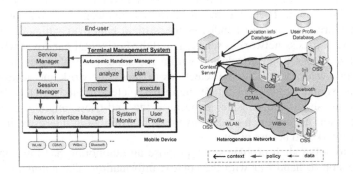

Fig. 3. Architecture of Terminal Management System using Autonomic Handover Manager

There are two major parts in this architecture; network side and terminal side. An optimal handover decision is to be achieved through their joint contributions. On the network side, the OSS of each network will perform network monitoring and report to the context server. The various repositories distributed in the networks will store the context information such as location information and user profiles. The context server located in the network collects the relevant context information from the context repositories. It prepares the evaluation matrix in response to the handover decision request from TMS. On the terminal side, the TMS will interact with the context server mentioned above for the purpose of making the optimal selection of the appropriate radio segment to which the terminal will eventually be assigned. The terminal's estimation of signal strength and QoS levels in the system are beneficially combined for making an informed selection of the appropriate radio technologies through which services can be obtained as efficiently as possible. Thus, both the network and the terminal contribute useful information towards the optimal decision.

AHM interacts with context server, system monitor and user profile repository. After processing the context evaluation, it decides on an appropriate policy and informs it to the service manager. The service manager manages the mapping of service and applications. It creates a session for each application using the policy from the AHM. The session manager maintains the created session and sends the decision of the appropriate network interface for the application to the network interface manager. The network interface manager serves two purposes. One is the retrieval of measurements at L2 level in the network interface and the other is the connection of the appropriate interface during a handover. Our management solution is located in the management layer and is independent of the OSI layers. Thus, it can control and monitor all layers. The system monitor collects the system information such as the remaining quantity of the battery and the memory. There are user settings on the services in the user profile.

AHM provides four major functions to decide an appropriate policy by the context information as mentioned in Section 2.

- Monitor function provides the mechanisms that collect, aggregate and filter the dynamic context from the terminal such as the received signal strength, cpu usage and remained battery.
- Analyze function provides the mechanisms that correlate and model complex situation. This updates the evaluation matrix from the context server by the monitored dynamic context.
- Plan function provides the mechanisms that construct the actions needed to achieve goals and objectives. This uses policy information to guide the result of evaluating matrix.
- Execute function provides the mechanisms that control the execution of a plan with considerations for dynamic updates. This publishes the policy to the service manager.

These four parts work together to provide the control loop functionality. The four parts communicate and collaborate with one another and exchange appropriate knowledge and data.

4.2 Context Evaluation Function

We show the context evaluation function to make optimal handover decisions based on the context information. More specifically, we need to select an appropriate network interface for each application considering the changing context information. This function requires a flexible data structure to express both user preferences and operator preferences. Thus, we define the context evaluation matrix which is flexible for updating.

Eq 1 shows the evaluation matrix for a service. A service is composed of m applications and the mobile terminal has n network interfaces. The k parameters are considered as well.

$$\begin{pmatrix} S_{11} & S_{12} & \cdots & S_{1n} \\ S_{21} & S_{22} & \cdots & S_{2n} \\ \vdots & \vdots & \ddots & \vdots \\ S_{m1} & S_{m2} & \cdots & S_{mn} \end{pmatrix} = \begin{pmatrix} C_{11} & C_{12} & \cdots & C_{1k} \\ C_{21} & C_{22} & \cdots & C_{2n} \\ \vdots & \vdots & \ddots & \vdots \\ C_{m1} & C_{m2} & \cdots & C_{mk} \end{pmatrix} \otimes \begin{pmatrix} R_{11} & R_{12} & \cdots & R_{1n} \\ R_{21} & R_{22} & \cdots & R_{2n} \\ \vdots & \vdots & \ddots & \vdots \\ R_{k1} & R_{k2} & \cdots & R_{kn} \end{pmatrix}$$

$$\Rightarrow (C_{i1} \quad C_{i2} \quad \cdots \quad C_{ik}) \otimes (R_{1j} \quad R_{2j} \quad \cdots \quad R_{kj})^T = f(C_{i1}R_{1j}, C_{i2}R_{2j}, \cdots, C_{ik}R_{kj})$$

$$(0 \le C_{ij} \le 1, 0 \le R_{ij} \le 1)$$

Eq. 1. Context evaluation matrix (m: # of applications, k: # of parameters, n: # of network interfaces)

- S_{ij} : The expected value of i application on using j network interface (i: index of application, j: index of network interface)
- C_{ij} : Coefficient (weight) of different parameters. This can be service specific (e.g. user preference) or service and operator specific (e.g., pricing) (i: index of application, j: index of parameter)
- R_{ij} : Parameters which will affect the decision (e.g., available bandwidth, signal strength) (i: index of parameter, j: index of network interface)

In addition to the matrix, additional rules or conditions are considered which assign values to parameters. For instance, there are some rules to set the upper or lower bounds. These rules have to be evaluated before evaluating the matrix. The C matrix is set by the user preferences. We propose a score function of $f(C_{i1}R_{1j}, C_{i2}R_{2j}, \cdots, C_{ik}R_{kj})$. This function should present the effective score for evaluation. For instance, it might be the SUM function as shown in Eq 2.

$$f(C_{i1}R_{1j}, C_{i2}R_{2j}, \cdots, C_{ik}R_{kj}) = \sum_{m=1}^{k} C_{im}R_{mj}$$

Eq. 2. The example of the score function, SUM function

As mentioned above, C matrix is a combination of applications and parameters and R matrix is a combination of parameters and networks. For example, let us consider the multimedia conferencing service (MCS), which is composed of audio application (A1), video application (A2) and data application (A3). The mobile terminal is able to access multiple networks such as CDMA (N1), WiBro (N2), IEEE 802.11b WLAN (N3), and Bluetooth (N4). In addition, we can consider context information, such as, available bandwidth, power consumption, signal strength, cost of service, velocity, QoS and so on. The parameters are defined by the combination of such context information. We define the parameters as Quality (P1), Price (P2), Velocity (P3) and Power (P4). The values of A and P are set by the user preferences. P1/N is made by available bandwidth, signal to noise ratio (SNR), bit error rate (BER) and class of quality. P2/N is made by cost of service, billing mechanism and class of quality. P3/N is made by the supported moving speed. P4/N is made by signal strength, CPU usage and LCD refresh rate. We can make R matrix (P x N). It emphasizes that the parameter is an important factor in the network. Then, we can make C matrix (A x P). It presents the weight of the parameter in the specific application.

After achieving the complete matrix of C and R, we can achieve S matrix. Eq 3 shows that we can obtain the map of application and network interface applying policies.

$$
\begin{pmatrix}
S_{11} & S_{12} & \cdots & S_{1n} \\
S_{21} & S_{22} & \cdots & S_{2n} \\
\vdots & \vdots & \ddots & \vdots \\
S_{m1} & S_{m2} & \cdots & S_{mn}
\end{pmatrix}
\xrightarrow{\text{policies}}
\begin{pmatrix}
S_{1j} \\
S_{2k} \\
\vdots \\
S_{ml}
\end{pmatrix}
\equiv
\begin{pmatrix}
A_1 : N_j \\
A_2 : N_k \\
\vdots \\
A_m : N_l
\end{pmatrix}
\rightarrow Q_i
$$

Eq. 3. The map of application and network interface by the evaluation matrix (m: # of applications, n: # of network interfaces, i: index of services)

- S_{ij} : result of a service (i: index of application, j: index of network interface)

- A_i : application (e.g., Audio, Video, Data) (i: index of application)

- N_i : network interface (e.g., WLAN, CDMA, WiBro, Bluetooth) (i: index of network interface)

- Q_i : Service (e.g., Multimedia conferencing service)

As a result of context processing, we can achieve Q matrix for all of services. We will describe the detailed example of the context evaluation matrix in the next section using a sample scenario.

Fig. 4. Context processing between the mobile node and the context server

Fig 4 shows the procedure for context information collection and processing between the mobile node and the context server. The mobile node sends a request to the context server to ask for an evaluation matrix. The context server collects the context information from the network and composes an evaluation matrix based on the received static context information from the mobile node as shown in Table 1. It fills in the known values of the parameters which can be drawn from the static context information in the network. Then, it sends the evaluation matrix to the mobile node and also sends updates of the parameters for dynamic context information in the network to the mobile node. The mobile node fills in the dynamic context information and calculates the evaluation matrix when a decision is needed. At this time, the context server should send the evaluation matrix when the dimension or component of the matrix is changed. Finally, the mobile node makes the decision based on the result matrix Q. A vertical handover is usually needed while the terminal is moving. Thus, the handover decision must be made very quickly within the specified time. Otherwise, the terminal will not be able to handover to the optimal network to provide continuous and seamless service.

5   Evaluation

In this section, we present an interesting scenario in 4G networks along with the evaluation result using our AHM to manage handovers with the goal of optimizing the selection of the new access network by the context information from the mobile terminal, the network and the user.

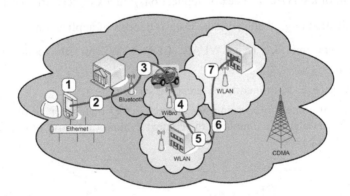

Fig. 5. Evaluation scenario (using multimedia conferencing service in 4G networks)

A business man is using the multimedia conferencing service on his PDA with a colleague in another city. He is initially connected from his home as in Fig 5. He prefers to use the service that is cost-effective, battery-effective and quality-effective. In order to satisfy his needs, TMS should decide the best network interface for him to use. AHM is able to provide the solution by evaluating the context information. The user can set the preferences of the application on cost, power and quality. After setting them once, the user does not wish to set them again while using the service.

Table 2. The result of AHM evaluation

APP	1	2	3	4	5	6	7
Available Networks	CDMA Wired	CDMA	CDMA Bluetooth	CDMA WiBro WLAN Bluetooth	CDMA WLAN	CDMA	CDMA WLAN
Audio	Wired	CDMA	Bluetooth	WiBro	CDMA	CDMA	WLAN
Video	Wired	X	Bluetooth	WiBro	WLAN	X	WLAN
Data	Wired	CDMA	Bluetooth	WiBro	WLAN	CDMA	WLAN

In Section 4, we have already explained the context evaluation matrix of the multimedia conferencing service. We have applied it to the above scenario. In Fig 5, we are able to make a decision for the vertical handover in 3, 4, 5 and 7. The user prefers to use the best quality audio application and video application and data application at the cheapest price. Table 2 shows the result of AHM evaluation for the vertical handover.

Even though there are same network choices in 5 and 7, the result is different because the network traffic load of 5 might be heavier than 7. We describe the process of the evaluation about 5 in Fig 5. In the 5th case, the available networks are WLAN and CDMA. The parameters are only quality and cost. First of all, AHM should make the context evaluation matrix by collecting the context information from the terminal and the network.

$$C : (Audio, Video, Data) \times (Quality, Cost), R : (Quality, Cost) \times (WLAN, CDMA)$$

$$f(C_{i1}R_{1j}, C_{i2}R_{2j}, \cdots, C_{ik}R_{kj}) = \sum_{m=1}^{k} C_{im}R_{mj}$$

$$\begin{pmatrix} 0.9 & 0.1 \\ 0 & 1.0 \\ 0.1 & 0.9 \end{pmatrix} \otimes \begin{pmatrix} 0.3 & 0.5 \\ 0.7 & 0.3 \end{pmatrix} = \begin{pmatrix} 0.34 & 0.48 \\ 0.7 & 0.3 \\ 0.66 & 0.32 \end{pmatrix} \xrightarrow{P} \begin{pmatrix} Audio : CDMA \\ Video : WLAN \\ Data : WLAN \end{pmatrix}$$

Eq. 4. The example of the context evaluation matrix

In Eq 4, the quality value of WLAN is 0.3 while that of CDMA is 0.5. This means that CDMA has a better quality because the network traffic load might be heavy in WLAN. Moreover, the result matrix shows that the audio application should select CDMA as the best network in this situation because the result of CDMA is greater than that of WLAN. In contrast, Video and Data applications should select WLAN. In this case, the calculation method of the normalized value of the matrix is an important aspect to be evaluated. However, this is not a concern for this paper. We assume that the other components of the terminal management system can explore this area.

6   Conclusion and Future Work

The handover management in 4G networks is a very important issue. We need a clever and seamless vertical handover mechanism when moving among various access networks. In this paper, we have proposed the AHM to provide a solution for determining the best network interface for the service. The decision is made by using the context information from the mobile node, networks and the user as well as the received signal strength. Thus, the end-users can enjoy efficient services according to their preferences in 4G networks. AHM is based on the autonomic computing concept. It provides a good policy for the vertical handover using the context information without user's interventions. We have presented the static and dynamic context information of the mobile node and the network. AHM has four major functions such as monitoring, analyzing, planning and executing. We have also described how to compose the context evaluation function and formulate a policy.

We have presented an interesting scenario in 4G networks to validate our approach, which is to use the multimedia conferencing service on the mobile terminal through the heterogeneous networks. The result has shown that the users are able to use flexible services according to their preferences on cost-effective policy, quality-effective policy and battery-effective policy.

For future work, we will present more concrete context information and improve the AHM by considering it and optimizing the context evaluation function. And, we will present more exact values for the evaluation matrix and analyze the overhead and cost associated with setting up the context server and transmitting the evaluation matrix periodically. Then, we will implement it on the open mobile platform and validate the feasibility. We will present how to apply the user preferences to the coefficient matrix of the context evaluation function easily and study the context management in 4G networks. Finally, we will test the various scenarios using the user profile data and network traffic analysis data to present the efficiency of our AHM.

References

[1] S. Balasubramaniam and J. Indulska, "Vertical Hanover Supporting Pervasive Computing in Future Wireless Networks," Computer Communication Journal, Special Issue on 4G/Future Wireless Networks, vol 27/8, pp. 708-719, 2003.

[2] E. Gustafsson and A. Jonsson, "Always Best Connected," IEEE Wireless Communications, vol 10, pp. 49-55, February 2003.

[3] IBM Corporation: An architectural blueprint for autonomic computing. White Paper, (2003).

[4] J.O. Kephart and D.M. Chess, "The Vision of Autonomic Computing," Computer, vol. 36, no. 1, pp. 41-52, Jan. 2003.

[5] H.J. Wang, R.H. Katz, and J. Giese, "Policy-Enabled Handoffs Across Heterogeneous Wireless Networks," Proc. of ACM WMCSA, 1999, p. 51.

[6] M. Angermann and J. Kammann, "Cost Metrics For Decision Problem In Wireless Ad Hoc Networking," in Proceedings IEEE CAS Workshop on Wireless Communications and Networking, (Pasadena, USA), September 2002.

[7] Chen L. –J., Sun T., Chen B., Rajendran V., and Gerla M., "A Smart Decision Model for Vertical Hanoff," In Proceedings 4th ANWIRE International Workshop on Wireless Internet and Research (ANWIRE 2004), Athens, Greece, 2004.

[8] Chen L. –J., Sun T., Chen B., Rajendran V., and Gerla M., "Universal Seamless Handoff Architecture in Wireless Overlay Networks," Technical Report 040012, UCLA CSD, 2004

[9] A. Calvagna and G. Di Modica, "A User-Centric Analysis of Vertical Handovers," ACM Wireless Mobile Applications And Services On WLAN Hotspots 2004(WMASH 2004), Philadelphia, PA, USA, pp. 137-146, 2004.

[10] A. Hasswa, N. Nasser and H.S. Hassanein, "Generic Vertical Handoff Decision Function for Heterogeneous Wireless Networks," IFIP Conference on Wireless and Optical Communications, pp. 239-243, March 2005.

[11] Q. Wei, C. Prehofer, "Context Management in Mobile Environment," In Proceedings 3th ANWIRE International Workshop on Wireless Internet and Research (ANWIRE 2003), November 13, Paris, France, 2003.

[12] Paulo Mendes, Christian Prehofer, Qing Wei, "Context Management with Programmable Mobile Networks," IEEE Computer Communications Workshop, October 20-21, Dana Point, California, USA, pp. 217-223, Oct. 2003.

Evaluation of an Intelligent Utility-Based Strategy for Dynamic Wireless Network Selection

Olga Ormond[1], Gabriel Miro-Muntean[2], and John Murphy[1]

[1] Performance Engineering Laboratory, School of Computer Science and Informatics,
University College Dublin (UCD), Dublin, Ireland
{j.murphy, olga.ormond}@ucd.ie
[2] Performance Engineering Laboratory, School of Electronic Engineering,
Dublin City University (DCU), Dublin, Ireland
munteang@eeng.dcu.ie

Abstract. In the case of next generation wireless networks, different users with various multi-homed personal wireless devices will aim to exploit the full potential of the choice of services and applications available over different radio access networks. In their selection of a certain radio access network consumers will consider money and delay preferences for the current application and will rely on intelligent network-selection decision strategies to aid them in their choice. This paper describes the evaluation of an intelligent utility-based strategy for network selection in a multi-access network situation for transfer of large non real-time data files. A number of scenarios are examined which compare the proposed network selection strategy against other possible strategies. Test results show how by using this network selection strategy significant benefits in terms of combined average delay and cost per file transferred as well as transfer efficiency are obtained.

Keywords: Heterogeneous Wireless Networks, Multi-homed Radio Terminals, User-centric Network Selection.

1 Introduction

Future generations of wireless networks will see the integration and interoperability of a range of current and emerging technologies, and with that an extensive collection of novel and attractive services will be produced by an array of service providers [1]. In order to take advantage of the choice of access networks and the variety of services together with the advances in portable computing, user devices will be built as multi-homed devices. The user-centric view for next generation networks is a service oriented heterogeneous wireless network environment (SOHWNE) envisioned as a setting where users and service providers will be free from subscription to any one network operator. They can instead choose the most suitable transport offering from the available network providers for their current mobile terminal and application requirements [2]. End-to-end sessions will be under intelligent terminal control with

A. Helmy et al. (Eds.): MMNS 2006, LNCS 4267, pp. 158–170, 2006.

the goal of always connecting to the right access network at the right time to service the current connectivity requirements according to the user preferences.

Section 2 of this paper describes the body of research carried out on the access network selection decision problem. Section 3 then presents the details of our proposed solution to the multi-access decision problem, a Consumer Surplus-based network selection strategy for non real-time data. Using predicted rates for the various available networks we find an intelligent decision strategy for maximising the users' cost savings, while meeting data delay constraints. The dynamic access network selection scenario for large files considered in this paper is presented in section 4. Section 5 outlines the testing scenarios in which the proposed strategy is compared against an Always Cheapest (AC) strategy, a Network Centric (NC) strategy, a random strategy, and a strategy from the literature. The paper ends in section 7 with conclusions and future work directions.

1.1 Motivation

A multi-mode terminal in a heterogeneous Radio Access Network (RAN) environment may be exposed to any combination of RANs (such as GPRS, UMTS, WLAN, WIMAX, Bluetooth, Satellite) at any time. Available bit rates on different technologies affect the types of service and the range of service quality levels that can be offered to users. The range of innovative data applications increases as the range of data rates improves. The intense competition in the marketplace will mean that both access and service providers are eager to survive and thrive by satisfying user demand for 24 hour ubiquitous access to online services. All providers are under pressure to supply the demand for user perceived quality services at the right price, meeting elevated bandwidth demands and supplying flexible services. These services need to be dynamically adaptable to the current context, such as the user terminal capabilities, preferences, and available access network characteristics.

The user-centric view is often referred to in the literature as the Always Best Connected (ABC) concept. Hotspots in airports, coffee shops and other public places already encourage mobile users with a dual terminal in an area of overlapping coverage to avail of the choice of their subscribed operator or a possibly cheaper public hotspot. Their choice will, of course, depend on a number of parameters based on application characteristics, monetary cost, user perceived quality of service, security and other networking concerns. As WLANs become more advanced, with established and secure billing systems they will increase in popularity. Users with private wireless home or office networks may also switch between these networks and the available public networks. However, the user cannot be expected to monitor these RANs, to assess the benefits from using these networks and switch manually to the RAN that best matches user expectations in terms of delay and cost. Therefore an intelligent and automatic solution to network selection is required.

2 Related Works

Much of the work in RAN selection policy design investigates the access network selection problem as part of the seamless handover venture. However, access selection is also relevant to the initial network selection decision for any user with a choice in available RANs.

Providing a positive user experience is crucial to the success of next generation networks, it is therefore important to take the user preferences into account in the access selection decision. Many of the RAN selection strategies do consider user requirements and preferences, but in the context of the network-centric vision. These solutions aim to describe the user preferences and then maximise the social welfare, or overall satisfaction, for all the users in a sole operator's network. They use the user preference information to the operator advantage to maximise revenue and network utilisation. For example in [3] Chan et al use their understanding of user behaviour to maximise network gains. They do not consider user-centric approach but rather look at congestion-based pricing to influence the user behaviour with the goal of optimal resource allocation.

A smaller portion of the existing literature consider the always best connected user-centric scenario where users take advantage of the operator competition and select the best network for current service in the current context. Examples are [4] and [5] both of which propose a network selection technique to provide the user with the current best available network, taking into account the user preferences, current available connectivity and the application requirements.

The architecture for implementing the network selection functionality is the focus of much of the work. In designing a new architecture, it is important to limit the modifications required to existing wireless systems, and to minimise the amount of additional network traffic needed. In [6] Eddy addresses the problem of which layer in the IP stack does mobility belong to. He analyses the disadvantages of Mobile IP (a network layer protocol) and concludes that while the Transport Layer is the most likely place for a mobility protocol, a cross-layer approach where interlayer communication is used may be the best approach.

In [7] Stemm et al, an architecture is proposed for vertical handover for multi-homed nodes in a wireless overlay network. This is based on Mobile IP but with some modifications. A Handoff Controller is implemented in the mobile terminal to make the decision of which network and which base station (BS) to connect to. This selection decision is based on network availability as indicated by increase or decrease in received beacons from the surrounding BSs, but it may be influenced by network constraints outlined by the user in the User Control Panel (on the terminal) or by heuristic advice from a subnet manager. The focus of their paper is reducing handoff latency; details on the user specified network constraints are not outlined.

Ylitalo et al in [8] look at *how* to facilitate a user making a network interface selection decision. They concentrate on a possible architecture for the end terminal and do not concentrate on any particular strategy but do mention the Always Cheapest (AC) network selection strategy. Their mechanism is based on Mobile IPv6 with a policy driver module located in the mobile terminal. The user defines their preferences via a GUI in the terminal. These preferences are then interpreted by the policy driver module together with the available information on all the network interfaces to determine a suitable set of actions.

Adamopoulou et al also present a mobile terminal architecture in [4]. They incorporate a GUI on the terminal to collect and weight four parameters (quality, preferred network operator, preferred technology type and cost) in order of the indicated user preference for each service in use. Their terminal management architecture consists of a network interface adaptation module, a mobility management module and the user preference module. The network selection is performed by the intelligent access selection function in the mobility management module. The cost information is received from the networks in advance of network selection and is not expected to change frequently. Each network operator charges per data volume, or per unit time, for access at a specific quality level. Multiple simultaneous applications are catered for with different networks selected for each application.

Song et al have their user-centric network selection module implemented in the link layer, with cross-layer signaling messages delivering the QoS information from different layers in the IP-stack, including the application layer where the user describes their desired attributes in a QoS Context. The scheme is based on a large number of parameters which describe; availability, throughput, timeliness, reliability, security, and cost. Two mathematical techniques, analytic hierarchy process and grey relational analysis, are used to perform the analysis and tradeoff between the parameters. The decision maker then chooses the network with the best score.

We focus on the user-centric decision making problem of *which* available network to choose for data transfer and focus on serving one applications communications needs at a time. The architecture is assumed to be one where the decision-making module is located in the mobile terminal and cross-layer information is available, but the contribution of the paper is the decision process itself and not the architecture.

In [9], H. J. Wang et al describe a handover 'policy' for heterogeneous wireless networks, which is used to select the 'best' available network and time for handover initiation. They consider the cost of using a particular network to be in terms of the sum of weighted functions of bandwidth, power consumption and cost. The network which is consistently calculated to have lowest cost is chosen as the target network. A randomised waiting scheme based on the impact of the estimated handover delay is used to achieve stability and load-balancing in the system, and to avoid handover synchronisation.

A number of papers, all of which reference [9], use similar cost functions. A smart decision model for vertical handoff is the focus of Chen et al in [10]. Their proposed scheme relies on a score function based on functions of allocated bandwidth (transfer completion time), battery power consumption and cost charged by the available networks.

The value or benefit function of offered link capacity is a concave increasing function following the economic assumption of diminishing marginal utility, i.e. value will increase for each unit of added bandwidth up to certain point after which the gain in value for extra capacity is marginal as described as in Murphy et al [11]. The bandwidth in this case is measured using a probing tool. The HTTP handoff decision model presented by Kammann et al in [12] and the work in [13] also use a cost function which is used to compare available networks from the list of options and establish the network to handoff to according to the importance weightings associated with different metrics.

When the wireless terminal seeks information on available access networks, the amount of information supplied should be minimal. Receiving current network condition information may be resource consuming and wasteful in the ever changing unreliable wireless environment. We use previous file delays experienced in a particular network to predict transfer completion time for that network.

In [9], bandwidth is determined either by use an agent in each RAN which estimates and broadcasts the current network load, or in the case of commercial networks, by the 'typical' value of bandwidth advertised by these networks. The bandwidth in [10] is measured using a probing tool. The main disadvantage of the probing technique is that it generates traffic adding an overhead in the different RANs. Both [9] and [10] collect current information on bandwidths available on all local networks – this requires heavy power consumption a factor which should possibly be added to the cost of implementing the suggested strategies. Also it is possible that the network becomes more congested just after the information has been collected.

The user network selection decision strategy will be influenced by the pricing scheme employed in the available networks. In our work we assume all available networks employ a price per byte pricing scheme, as i-mode users are currently charged. Our scheme could also cope with other pricing schemes where the cost is known in advance of making the decision, and will not change for the duration of the session.

3   Consumer Surplus Network Selection Strategy

In our work we endeavor to find an intelligent solution to the selection decision problem for non real-time data services. In the case of large data transfers, we believe that an initial correct decision may save the user the effort of handoff later in the call session. This decision will rely heavily on the charging schemes implemented by the candidate networks, as users make an effort to get the best value for their money. This paper considers the case when a user wishing to send non real-time data (a large file) has a choice of several available networks, each of which employ a fixed price per byte pricing scheme but each charging different prices. Every user wants timely delivery of their data at the lowest price. While the strict time restrictions imposed by real-time traffic are not used here, it is assumed that every user has a patience limit and will only be willing to wait so long for completed transfer of their data before they become dissatisfied.

We employ a non increasing user utility function to describe user's willingness-to-pay (in cent) in relation to their willingness-to-wait (in seconds) for a particular service transfer. The more delay a user experiences the less they may be willing to pay. We define two thresholds in terms of time to transfer completion. Any completion time up to T_{c1} is worth the highest price to the user, whereas files arriving after T_{c2} are worth 0 cent.

In our work in [14], we explored possible shapes for the user utility function. Previous work in the field of economics had related three different utility function shapes with three different users' attitudes to risk taking. The user utility function used in this paper is related to risk neutral user behaviour, where the user equally values paying less money to waiting shorter transfer completion times. It is described in (1) below.

$$U_i(T_c) = \begin{cases} U_{\max,} & T_c \leq T_{c1} \\ U_{\text{var}} - T_c, & T_{c1} < T_c \leq T_{c2} \\ 0, & T_c > T_{c2} \end{cases} \tag{1}$$

The difference between the value of the data to the user (willingness-to-pay) and the actual price they are charged is known in microeconomic terms as the Consumer Surplus (CS). The consumer surplus for sending a file i (**CS$_i$**) is described in equation (2) below, where **U$_i$(T$_c$)** is the utility value in cent of transferring file i in time **T$_c$** seconds and **C$_i$** is the cost charged in cent for sending the file via the chosen network.

$$CS_i = U_i(T_c) - C_i$$

$$s.t. \ T_c \leq T_{c2} \tag{2}$$

In this strategy we use the file size and throughput predictions to predict the transfer completion time. Together with the utility function, this will give us a predicted utility value for each network, which in turn is used together with each network costs to get the predicted consumer surplus. We then choose the network with the best predicted CS.

There are different possible network selection strategies for the user. A user may choose to always stick with one particular network regardless of its current characteristics, for example to always select the cheapest network, or to always chose their preferred operator's network (a network-centric strategy). Another possible strategy would be a random network selection. None of these strategies are responsive to changes in the dynamic radio environment, and in the case that the chosen network is heavily congested the user will be stuck with a poor network choice. However, we believe that an intelligent selection based on user willingness-to-pay, their file completion time constraints and estimated access network delivery time should be used instead. These are considered by our proposed CS-based network selection strategy.

In this paper, a large file is fragmented into smaller chunks and a network selection decision is made for each chunk. The objective of the decision is to transfer the entire file (chunk by chunk) over the current best of the available access networks. It uses the throughput rate information from previous chunks to predict the available rates in each network. *We propose a utility-based algorithm that accounts for user time constraints, estimates complete chunk delivery time (for each of the available access networks) and then selects the most promising access network based on consumer surplus difference.*

The scenario considered here is where all the network operators employ a fixed pricing scheme. The weighted average of the throughput rates experienced for the last five chunk transactions in a particular network is used as the predicted rate for that network. In the case where there are no previous rates stored for a given network a default rate is used (that is the minimum acceptable rate which will meet the user's

T_{c2}). If the predicted rate for a network is bad then this low rate is only used temporarily before it is deemed stale information and replaced by the default rate.

4   Access Network Selection Scenario

An intelligent mobile user's FTP application requires the transfer of a large file (size 100Kbytes - 5Mbytes) in a SOHWNE. This is a typical range for the size of an FTP transfer. The user terminal has inbuilt intelligent features for network detection and selection. The file is broken into smaller chunks of data (length 10Kbytes). Each data chunk is to be sent uplink from the terminal through the access point (AP) of the selected network to a server in the wired network. The wired network is connected via point-to-point links so that delay in the wired network is negligible and any significant delays experienced are dependent on the selected radio access network. The user may for example be uploading a photo slideshow from a laptop, emailing a large document from their mobile phone, or uploading vending machine data from their PDA to a central server.

The user is faced with a scenario like the one depicted in Figure 1, where the terminal employing the intelligent network selection strategy has a choice of two IEEE 802.11b networks: WLAN0 or WLAN1. We consider that the transfer over WLAN1 is twice as expensive as WLAN0. A repeated decision is required for which RAN to use for transporting each chunk of user's application data.

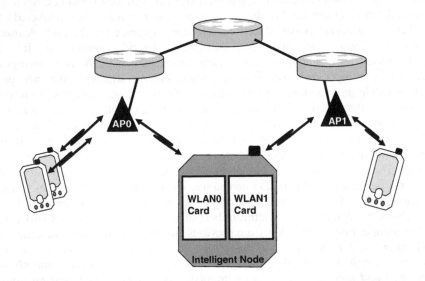

Fig. 1. Simulation scenario setup: user device has a choice of two IEEE 802.11b networks: WLAN0 or WLAN1

The simulation model was developed in NS2 version 2.27 [12] with IEEE 802.11b wireless LAN parameter settings (data rate 11 Mbps), the NOAH (No Ad-hoc routing) extension, and our new application which simulates a multi-homed terminal with a set of inbuilt network selection strategies. Five decision strategies are supported:

- Consumer Surplus Strategy (CS)
- Always Cheapest Strategy (AC)
- Network Centric Strategy (NC)
- Random Strategy (R)
- Another intelligent strategy as proposed in [9] (WKI)

CS network selection strategy conforms to the description presented in section 3. AS network selection strategy aims at minimizing the cost of data transfer, regardless of the performance of the transmission, selecting the cheapest network available (in this situation WLAN0). NC network selection strategy considers the case when the transfer is always performed over the network owned by the user's preferred network operator (in this case WLAN1 - the more expensive network). The random network selection, randomly selects either WLAN0 or WLAN1. The Wang-Katz I method (WKI) is an implementation of [9] using their formula (3). We implemented the formula with the weightings for cost and bandwidth both set at 0.5, the weighting for power consumption set to 0, and used our own throughput prediction scheme. We choose these weightings which convey the situation where the user equally values data delay to the price of the data transfer. The user utility function we used for our CS strategy models the same user behaviour.

$$f_n = w_b \bullet \ln \frac{1}{B_n} + w_p \bullet \ln P_n + w_c \bullet \ln C_n$$

$$s.t. \sum_i w_i = 1$$

(3)

The quality of the intelligent node's connectivity in each RAN is affected by the background traffic caused by the other nodes in the cell. In IEEE 802.11b networks if the bulk of data traffic is on the downlink, an FTP session transferring data uplink would have a clear advantage as the bulk of the traffic would create a bottleneck at the AP. The background traffic mix used in this simulation was a mix of HTTP, WAP, FTP and two-way non real time video sessions. The connectivity quality is affected mainly by the background FTP and video traffic. Although WAP is not a common application in current WLANs, in the future when smart-phone users will be taking advantage of the cheaper rates on a WLAN they may be only interested in WAP sessions on their smaller screens.

5 Comparison-Based Performance Evaluation

During successive tests file size is varied from 100Kbytes to 5Mbytes. All five network selection strategies: CS, AC, NC, R and WKI were employed in turn. In the case of CS network selection strategy the intelligent node will shortlist the networks based on delivery time prediction to meet T_{c2}. Each chunk is then transmitted over the short-listed RAN which is predicted to maximise the CS. If none of the available networks can meet the delivery deadline, then the cheapest network is selected.

If a chunk transfer exceeds the transfer completion deadline, the transmission is stopped and the chunk is resent. The entire file transfer completion time, cost of transfer, number of chunk resends required and consumer surplus are then compared against two other network selection strategies: AC and NC. The background traffic pattern on WLAN0 changes from high traffic to medium whereas WLAN1 starts with a medium amount of traffic which increases. As WLAN1 is twice as expensive as WLAN0 we assume that it will get slightly less traffic and allowed for that in the background traffic patterns.

In the experiments, it was noted that the WKI strategy only ever selected the cheapest network, WLAN0. This was due to the default setting for our throughput prediction scheme. As WLAN1 was twice as expensive as WLAN0, it needed to be twice as fast in order to be selected under the WKI scheme. With our default throughput setting, which was low, WKI strategy never selected WLAN1 and so the actual throughput rates available on WLAN1 were not identified. As a result of this, we changed the default setting to a higher bandwidth and re-ran tests for 5Mbyte files under the strategy name WKII. While WKII compares favourably with our strategy on the average transfer completion time, it does not fair as well on the chunk resends and the average cost per file.

5.1 Average Delay Per File Transfer

The average transfer completion delay per file (measured in seconds) is compared for the all the network selection strategies and the results are plotted in figure 2. The

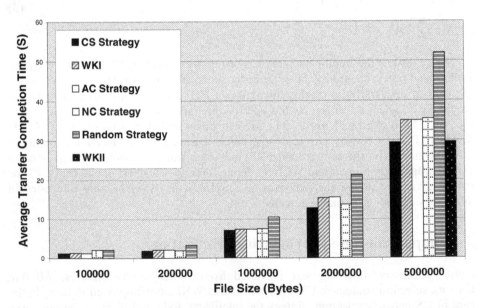

Fig. 2. Average File Transfer Completion Time

transfer completion time is important to the non real-time users as they are not willing to wait too long for the file transfer to finish. The sooner the file transfer ends, the more satisfied the user is. Analyzing the results it can be seen that the average duration for files sent with the CS-based strategy are the shorter than files sent with any of the other methods, with the exception of WKII. As file size increases the difference between the CS average transfer time and the others increases. For example when transferring 5000000 byte files using CS strategy, 17% shorter times for transfers are recorded than when AC strategy is employed and 19% shorter than when NC approach is used.

5.2 Average Cost Per File Transfer

The user wants to take advantage of the competition and pay as little as possible to achieve the expected perceived quality for the current service. Figure 3 shows the comparison of average cost (in cent) per file transfer using each of the three strategies considered in this paper. As expected with the NC strategy, users who always choose their network operator's WLAN (WLAN1) end up paying the most as this is the most expensive network of the two. For the case of transferring the maximum file size considered in this paper, users on the NC scheme pay on average double what users on the AC scheme pay. Although as expected the AC strategy achieves the least costly transfers, the difference between the average costs paid per file for the CS and AC strategy users is not very much, compared to the difference in average delay.

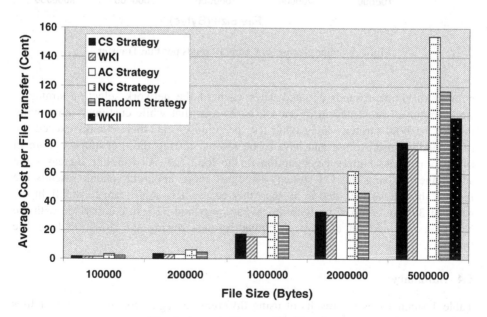

Fig. 3. Average Cost per File Transfer

5.3 Transmission Efficiency Per File Transfer

One measurement of transmission efficiency is in terms of how many network resources were used in order to transfer a given amount of data. Looking at the

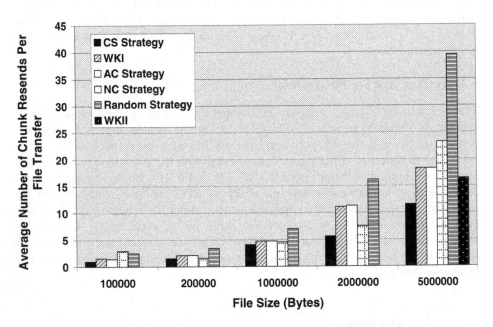

Fig. 4. Average Number of Chunk Resends per File Transfer

number of chunk resends counted when each of the network selection strategies were employed presented in figure 4 it is obvious that while customers employing the AC or WKI strategy are paying less per chunk sent, their chunks exceed the time deadline more often and have to be resent, costing these customers more in time, money and battery consumption in the long run. As file size increases the advantage of using the CS strategy becomes more apparent, chunk resends are far less than those recorded in all the other cases. It is significant to highlight the gains of 39% in comparison with the AC strategy and 52% in comparison with the NC plan obtained on average in chunk resends during the 5000000 byte file transfers.

5.4 Summary

Table 1 summarizes the results of using different strategies for the transfer of files with various sizes.

Table 1. Comparison between CS, AC and NC strategies when transferring different size files

File Size	Strategy	Average Duration (seconds)	Average Cost (cents)	Average Resends (times)
100000	CS	1.184	1.782	0.980
	AC	1.233	1.609	1.417
	NC	2.133	3.419	2.916
200000	CS	1.882	3.661	1.528
	AC	2.130	3.142	2.117
	NC	1.925	6.275	1.610
1000000	CS	7.045	17.357	4.072
	AC	7.337	15.465	4.763
	NC	7.486	30.822	4.580
2000000	CS	12.833	32.992	5.604
	AC	15.551	31.035	11.327
	NC	13.746	61.527	7.529
5000000	CS	29.444	81.304	11.556
	AC	35.040	76.938	18.280
	NC	35.406	154.152	23.250
	WKII	29.579	98.522	16.464

6 Conclusions and Future Work

Thanks to deregulation and the advancements in technologies, the user is becoming more powerful in their everyday consumer choice. Users in a user-centric SOHWNE will be free to choose the available access network which can best satisfy their needs in their current situation. This paper proposes, describes and evaluates a novel algorithm for intelligent cost-oriented and performance-aware selection between available networks. This utility-based model for user network selection divides large files into smaller chunks and selects the best available network for transferring each chunk, with user-specified time constraints, in a heterogeneous wireless environment.

The ultimate goal is to find the best user-centric network selection strategy for non real-time data transfer in the next generation service oriented heterogeneous wireless network environment. In this paper the proposed Consumer Surplus (CS)-based strategy was evaluated against an Always Cheapest (AC), a Network Centric (NC), a Random and two forms of an intelligent selection strategy (WKI & WKII). Test results showed that of all the comparison strategies WKII was the strongest contender.

WKII compares favourably with our CS strategy on the average transfer completion time, but it is not quite good as on the chunk resends and the average cost per file transfer.

More rigorous testing of the CS in compassion with forms of the WK strategy are currently underway, including scenarios with different competing traffic mixes and different chunk sizes.

Future models will use more advanced throughput prediction methods that consider the time of day, day of week, previous history, congestion recovery time and other possible influential metrics.

Acknowledgments. The support of the Informatics Commercialisation initiative of Enterprise Ireland is gratefully acknowledged.

References

1. Houssos, N., Gazis, V., Panagiotakis, S., Gessler, S., Schuelke, A., Quesnel, S.: Value Added Service Management in 3G Networks. IEEE/IFIP Network Operations and Management Symposium (2002)
2. Kanter, T.: Going Wireless: Enabling an Adaptive and Extensible Environment. ACM Journal on Mobile Networks and Applications, Vol. 8, No. 1 (2003) 37-50
3. Chan, H., Fan, P., Cao, Z.: A Utility-Based Network Selection Scheme for Multiple Services in Heterogeneous Networks. IEEE International Conference on Wireless Networks, Communications and Mobile Computing (2005)
4. Adamopoulou, E., Demestichas, K., Koutsorodi, A., Theologou, M.: Intelligent Access Network Selection in Heterogeneous Networks - Simulation Results. IEEE Wireless Communication Systems Symposium (2005) 279-283
5. Song, Q., Jamalipour, A.: An adaptive quality-of-service network selection mechanism for heterogeneous mobile networks. IEEE Wireless Communications and Mobile Computing (2005) 697-708
6. Eddy, W. M.: At What Layer Does Mobility Belong? IEEE Communications Magazine (2004)
7. Stemm M., Katz, R.H.: Vertical handoffs in wireless overlay networks. Mobile Networks and Applications 3 (1998)
8. Ylitalo, J., Jokikyyny, T., Kauppinen, T., Tuominen, A. J., Laine, J.: Dynamic Network Interface Selection in Multihomed Mobile Hosts. 36th Hawaii International Conference on System Sciences (2003)
9. Wang, H.J., Katz, R.H., Giese, J.: Policy-enabled handoffs across heterogeneous wireless networks. Second IEEE Workshop on Mobile Computing Systems and Applications (1999)
10. Chen, L-J., Sun, T., Chen, B., Rajendran, V., Gerla, M.: A Smart Decision Model for Vertical Handoff. 4th ANWIRE International Workshop on Wireless Internet and Reconfigurability (2004)
11. Murphy, J., Murphy, L.: Bandwidth Allocation By Pricing In ATM Networks. IFIP Transactions, C-24, pp. 333-351, Broadband Communications II (1994)
12. Angermann, M., Kammann, J.: Cost Metrics For Decision Problem In Wireless Ad Hoc Networking. IEEE CAS Workshop on Wireless Communications and Networking (2002)
13. McNair, J., Zhu, F.: Vertical Handoffs in Fourth-Generation Multinetwork Environments. IEEE Wireless Communications (2004) 8-15
14. Ormond, O., Muntean, G.-M., Murphy, J.: Utility-based Intelligent Network Selection in Beyond 3G Systems. IEEE International Conference on Communications (2006)
15. Network Simulator-2, [Online]. Available on http://www.isi.edu/nsnam/ns/

Overlays on Wireless Mesh Networks: Implementation and Cross-Layer Searching

Gang Ding[1], John Vicente[2,3], Sanjay Rungta[3], Dilip Krishnaswamy[4],
Winson Chan[3], and Kai Miao[3]

[1] School of Electrical and Computer Engineering, Purdue University
[2] Department of Electrical Engineering, Columbia University
[3] Intel Information Technology Research, Intel Corporation
[4] Intel Mobility Group, Intel Corporation
dingg@purdue.edu, {john.vicente, sanjay.rungta,
dilip.krishnaswamy, winson.c.chan, kai.miao}@intel.com

Abstract. After years of research on ad hoc networks, practical wireless mesh networks are moving towards mainstream industry deployment. As wireless mesh networks become more ubiquitous, how to enable distributed applications and services is a challenging research topic. A new network architecture called OverMesh is recently proposed, in which computational overlays provide the facility to deploy distributed services across mobile mesh nodes. In this paper, we present the first implementation of the OverMesh architecture. The overlays are built over an IEEE 802.11s wireless mesh network pre-standard prototype. The platform enables development and deployment of concurrent distributed experiments on wireless mesh networks. Based on this platform, we further introduce a cross-layer searching algorithm, which combines traditional overlay searching with ad hoc network routing so that a physically short searching route is facilitated. Both experimentation and simulation results are presented.

Keywords: Overlay, virtualization, wireless mesh network, cross-layer.

1 Introduction

There has been a long history of research on mobile ad hoc networks in which peer nodes relay packets for other nodes while no fixed infrastructure exists to control or manage the network [2]. While wireless mesh networks have been actively studied [3], there has been increasing interest from the industry for its deployment use. A wireless mesh network is a form of wireless ad hoc network where some mesh nodes are stationary, thereby reducing the power constraint concerns. Standardization of wireless mesh networks for different settings have already started, for example, the IEEE 802.11s mesh network for wireless local area network (WLAN) [4], the IEEE 802.15.5 mesh for wireless personal area network (WPAN) [5], the ZigBee mesh network for low rate WPAN or sensor networks [6], and the mesh mode for IEEE 802.16 wireless metropolitan area network (WiMAX) [7]. Moreover, the next generation wireless network is expected to be a hybrid of all these mesh networks,

A. Helmy et al. (Eds.): MMNS 2006, LNCS 4267, pp. 171–182, 2006.
© IFIP International Federation for Information Processing 2006

which can be further integrated to the cellular networks and the Internet. When this vision becomes true in the near future, it is anticipated that a growing number of services and applications will run on it. While current research activities on wireless mesh networks are mostly focused on the physical, data link, and network layers; in this paper, however, we take one step further towards enabling distributed services and applications on the wireless mesh networks.

An obvious indication of the success of the Internet is the wide deployment of services and applications, including client-server applications such as the World Wide Web (WWW), email, and ftp, and more recently, the overlay networks and peer-to-peer applications. When we study future services and applications in wireless mesh networks, the overlay network would be preferred because it shares similar features with underlying wireless ad hoc networks, such as no centralized server, dynamic network topology, and localized operation. This motivates our interest to investigate an architecture of overlays on wireless mesh networks, called OverMesh [1]. However, the challenges to integrate them were not straightforward because the overlay network is a virtual network running in the application layer and the underlying network is transparent to the nodes in the overlay, while nodes in a wireless mesh network should participate in the actual routing in network or link layer. In addition, the mesh nodes in wireless networks are limited by bandwidth, computation capacity, and interferences, which is not a big concern in overlay networks on the Internet because of the high speed cable and dedicated routers. Therefore, how to enable efficient overlay services and applications on the resource constrained wireless mesh networks is a challenging research proposition. We believe that the cross-layer approach should be employed because it can facilitate information exchange between the overlay in the application layer and the network and lower layers. Therefore, we propose a cross-layer overlay searching algorithm which takes advantage of the broadcast based routing in the underlying wireless mesh network to quickly find the shortest physical searching route. The cross-layer searching protocol is realized and compared with the distributed hash table (DHT) [10] based overlay searching algorithms.

The rest of the paper is organized as follows. Section 2 briefly introduces the OverMesh architecture. Our implementation experience along with the details of the current OverMesh platform are described in Section 3. The cross-layer searching protocol is presented in Section 4. Section 5 gives the experimentation and simulation results. Related work is reviewed in Section 6. Section 7 concludes the paper and proposes some future directions.

2 OverMesh Networks

We have proposed a generic overlay architecture called OverMesh [1]. We view it as a parallel edge/access internetworking strategy positioned for novel use scenarios including, for example, residential or local community networks, office networks, home networks, and first-response networks. The followings are collectively the differentiating properties of OverMesh against traditional networking, peer-to-peer computing systems, or ad hoc networking systems:

Infrastructure-free. We position an aggressive convergence strategy for node computation, network processing, and data storage. Any physical node may be capable of supporting alternative properties of interconnection and source and sink functions;

Network virtualization. Facilitated by distributed virtual machines, the overlays would enable a computational model for provisioning and managing network structure and resources, distributed network services and applications;

Emergent control and manageability. To achieve the level of robustness and resilience seen in today's internetworking systems under a decentralized networking system, we argue that it is necessary that some services exhibit behaviors typically seen in emergent biological systems; we call for alternative learning and statistical inference techniques to off-load human-dependency on operational management;

Cooperative and adaptive end-to-end control. To support a horizontal and vertical system orientation to scale and adapt wireless communications, we believe the end-to-end principle and in-network control must converge. We call for tighter layer integration and cross-layer control and management.

The conceptual OverMesh architecture can be applied to a variety of wireless mesh networks. At its current stage, we chose to realize it on one of the mesh networks that are being actively standardized – the IEEE 802.11s WLAN mesh network [4]. The PlanetLab [9] architecture was customized and integrated with the WLAN mesh network to manage the distributed virtual machine based overlays. Overlays and virtualization have been proven to facilitate deployment of large distributed services and peer-to-peer applications on the Internet, but when applying them to the resource constrained mobile wireless environments, we pursed investigation of the following new issues:

Testbed for distributed applications on wireless networks. The testbed should provide an open platform for research on real wireless mesh networks. Researchers can develop and deploy their own distributed experiments and test the performance on the whole mesh network. Each experiment spans multiple mesh nodes. The virtualized overlay enables concurrent but isolated experiments on the same physical testbed. The details of realizing such a testbed is the focus of next section.

Efficient management of resource-limited wireless mesh networks. The traditional overlay for wired networks tries to make the underlying network transparent to the users. This may not be desirable for wireless networks. Given limited bandwidth, computation capacity, and dynamic topology in wireless networks, the resource management and control should consider the underlying network condition. Theoretical approach to the resource management in wireless ad hoc networks has been extensively studied. But the current wireless ad hoc networks still lack a generic platform to enable these results. The OverMesh architecture provides a feasible way to support distributed resource management and control: each node contributes one of its virtual machines to form a resource management overlay. All virtual machines in this overlay work together and try to balance the network load in a fully distributed fashion. The difference of this overlay from other overlays is that it can collect information from the network and physical layer or send control signals down to lower layers. This shows the importance of information exchange between different layers. In particular, this paper proposes a cross-layer searching algorithm in Section 4. It combines the overlay searching and the ad hoc routing so that a searching request from the overlay can be quickly replied with the least use of resources in the lower wireless network.

Develop distributed services for future wireless mesh networks. In addition to the large number of services already available in the wired network, there will be many novel services and applications specific to the next generation wireless networks, for example, distributed network measurement and monitoring, localization of mobile nodes, data aggregation and mining. This paper is not focused on a particular application, but we have implemented the open platform and will demonstrate how to develop and deploy a service on it. Results from extensive experimentations and simulations are reported in Section 5. We believe that the implemented OverMesh platform will provide a unique testbed for developing a wide variety of novel services and applications on wireless mesh networks.

3 Implementation

The prototype of wireless mesh network is based on the draft of IEEE 802.11s working group, called Simple Efficient Extensible (SEE) mesh. A mesh node provides 802.11 conformant MAC and PHY interfaces. Some mesh nodes serve as access points providing additional basic service set to support communications with simple mobile wireless clients. A mesh portal is a mesh node that specifically serves as an entry or exit point for packets in the network and routes packets into or out of the mesh network from other parts of a distribution service or non-802.11 networks. The implementation of 802.11s pre-standard is illustrated in Figure 1. It supports metric-based multi-hop routing and data forwarding [11] at link layer, neighbor discovery, link quality measurement, media access coordination with quality of

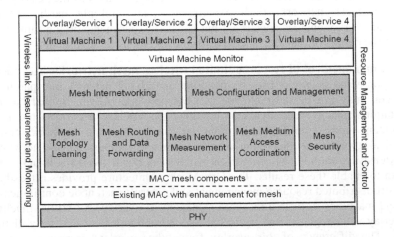

Fig. 1. Mesh node implementation

service support, security, and a user interface for mesh network configuration and management. To realize these mesh components, the firmware of the wireless network interface card is updated, new kernel modules and user interface applications are developed. The mesh prototype supports both Linux and Windows. However, all experiments introduced in this paper are conducted on Linux machines with Linux kernel 2.6.5+.

We employ the PlanetLab architecture to facilitate overlay maintenance and virtualization on the OverMesh platform. PlanetLab is a large distributed testbed for overlay networks on the Internet. It has evolved rapidly over the past three years and as of this writing, it encompasses 670 servers hosted at 326 sites and spanning 35 countries. PlanetLab employs Linux VServer [12] to virtualize servers. A central goal of PlanetLab is to support distributed virtualization – allocating a widely distributed set of virtual machines for a user community or application. Research groups are able to build their own virtualized overlay (called *slice* in PlanetLab) in which they can experiment with a variety of planetary-scale services, including file sharing and network-embedded storage, content distribution networks, routing and multicast overlays, scalable object location, anomaly detection mechanisms, and network measurement tools. Researchers continue to expand PlanetLab and investigate new research directions, such as introducing clusters to PlanetLab, using other machine virtualization technologies [23-27], federating PlanetLab, and private PlanetLab [34-36].

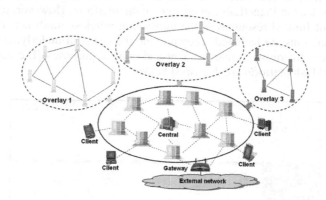

Fig. 2. OverMesh distributed system

As illustrated in Figure 2, the current OverMesh prototype includes the following four components:

Central is responsible for overlay management. Any new node joining the system needs to contact the Central in order to be authorized. Whenever a node is selected to participate in an overlay, its information is stored in the Central. But the Central itself does not need to participate in any distributed services provided by overlays, nor does it need to help with multi-hop routing in the wireless mesh network.

Gateways connect the mesh network to external networks, such as the Internet or other wireless networks.

Mesh nodes are the physical nodes participating in the multi-hop wireless communication and the overlay services, as represented by the computers in the solid circle in Figure 2. Each mesh node is partitioned to multiple separated virtual machines at the operating system level. Each virtual machine may participate in an overlay (dashed circle in Figure 2) that involves virtual machines from multiple mesh nodes. All virtual machines in an overlay form a virtual network where nodes collaborate with each other in order to provide a distributed service for any of the mesh nodes and clients in the system requiring access to the service. We assume that mesh nodes are mostly stationary; however, some may be mobile albeit maintaining a certain degree of operational stability and connectivity to the OverMesh system.

Clients connect to the nearest mesh node through wired or wireless links. They can participate in the multi-hop wireless networking and data forwarding if they are mobile and communicating through wireless links, but they do not contribute to providing overlay services. Thus they are not managed by the Central.

While PlanetLab targets large distributed overlay networks supported by dedicated servers in the Internet, we focus on realizing the virtualized overlay on personal computers in a wireless mesh network. By participating in the research activity on private PlanetLab [34], firstly we reengineered the existing PlanetLab service architecture to operate in a local intranet environment. We then simplified and customized the PlanetLab codes to operate on a private WLAN mesh network. Figure 1 shows the current system stack of a mesh node. In addition to the mesh networking components residing in link layer, the virtual machines are supported by a virtual machine monitor. It sits below the traditional network layer while the existing distributed services in PlanetLab can operate on the platform. However, to make more efficient use of limited resources in the underlying wireless mesh networks, several cross-layer functions are included. They provide the status of underlying networks to the upper layers for resource management and control purposes. The major steps to setup an OverMesh platform are described below.

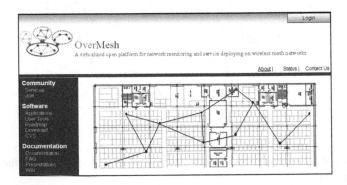

Fig. 3. Website at Central

1. Setup the Central. Although a server is usually used to host the Central in a private PlanetLab, we use a laptop as a Central in order to facilitate its wireless communication with other mesh nodes. It has an Intel Pro Wireless 2200 card [13] supporting IEEE 802.11b and g. It is connected to the external Internet through a cable so that it can download packages from external server. It also acts as a gateway so that all mesh nodes can also be connected to the Internet. The Central maintains the authorization, installation, and status of all mesh nodes. For this purpose, it hosts a web site for managing the mesh nodes, distributed virtual machines, and overlays. A snapshot of the homepage is displayed in Figure 3. It illustrates a possible mesh network topology in an office building with many cubicles and meeting rooms. The dots represent the Central and mesh nodes. They are all communicating through multi-hop wireless links.

2. Install a mesh node. We use laptops as mesh nodes. A new mesh node is booted from a CD which temporarily installs a light-weight Linux kernel in the memory and enables the wireless mesh networking. The CD contains a specific key file generated by the Central for each new node. When the new mesh node is booted from the CD, it first contacts the Central through wireless links to verify the key. If it is authorized by the Central, the new mesh node will continue to download a latest installation kernel from the Central and install it. In this manner, the installation kernel can always be upgraded by the Central. The mesh networking support was added to both the boot CD and the installation Linux kernel in the Central so that any new mesh node can always communicate with the Central through wireless links.

3. Setup the clients. A client only needs to install the mesh network software so that it can communicate to any node in the system through a nearby mesh node or client.

4. Deploy a service. When users want to deploy a new service on the mesh network, they should first request the administrator to add a new overlay through the website at the Central. Then a user can login to the website to manage this overlay. Since the Central has both wired and wireless connections, the user can login from either the external Internet or a mesh node or client in the mesh network. The users will be able to add or delete any mesh node in their own overlay. Every selected mesh node will be automatically notified and configure itself to add a new virtual machine to participate in the overlay. Finally, users can login to every participating virtual machine and install their particular service and application software without interaction with any other services supported by the same physical machine. Most existing services running on the public PlanetLab can be easily ported to this platform for wireless mesh networks.

4 Cross-Layer Searching

When realizing the OverMesh platform, we noticed that the existing overlay networks designed for the Internet do not work as efficiently on wireless networks as expected. One could experience significant delays and high packet loss rate. This is mainly due to the separation between the upper overlays and the lower network, MAC, and physical layers. How to provide fast and efficient services on the resource constrained wireless mesh networks is a challenging research topic. Current work on overlays usually does not consider the underlying network conditions. On the other hand, current research on wireless mesh networks is mainly focused on the network and lower layers. Our approach to deal with this issue is cross-layer adaptation. Cross-layer design for mobile wireless networks is becoming increasingly important [16, 17]. In general, it can be used in network monitoring and network resource management and control. Specifically, we study a cross-layer overlay searching algorithm in this paper.

A searching overlay, such as OpenDHT [14, 15] deployed in PlanetLab, provides a common lookup service to various applications such as information queries, distributed file storage and sharing. A (key, value) pair is stored in a randomly selected node in the searching overlay. However, it can be found by any node in the overlay using a searching algorithm. The most efficient overlay searching algorithms, such as Chord, Pastry, Tapestry, and CAN, are based on the DHT [10]. Each virtual node maintains a small overlay routing table. For Chord, Pastry and Tapestry, the routing table size is $O(\log n)$ and the hop count of a searching route is $O(\log n)$, where n is the network size. For CAN, the routing table size is $O(d)$ and the hop count of a searching path is $O(dn^{1/d})$, where d is a constant. However, these overlay searching algorithms only achieve the short searching route in terms of the number of virtual hops in the overlay, while they attempt to make the underlying network transparent to the nodes in the overlay. Figure 4a illustrates the overlay searching on the Internet. We assume that a (key, value) pair is stored in node D, while node A only knows the key and sends out a query for the corresponding value. The actual searching path of the query involves two loops. The outer loop is the searching in the overlay, the short virtual path A→B→C→D in the overlay can be found by some DHT based overlay searching algorithm. The inner loop is the real network routing. For example, in order to send the query from A to B, it should be routed in the physical network. But the overlay neighbors A and B may be physically far away from each other. In a wired network, the overhead in the underlying network is not a big concern due to the high speed cable and dedicated routers. However, for the multi-hop wireless mesh network, in order to find a physical route for each overlay hop, the source node has to broadcast a route request to all its neighbors repeatedly until the destination is reached, as demonstrated by Figure 4b. This makes the previous two-loop overlay searching algorithm on wired networks inefficient on wireless mesh networks.

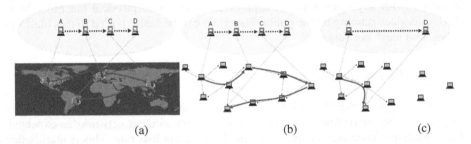

(a)   (b)   (c)

Fig. 4. Searching route on wired and wireless networks. (a) Overlay searching on the Internet. (b) Overlay searching on a wireless mesh network. (c) Cross-layer searching on a wireless mesh network.

Given factors such as the limited bandwidth available, sharing of the medium, power constraint, mobility, dynamic topologies, and varying link conditions in wireless mesh networks, the searching algorithm should consider the network condition in lower layers in order to quickly find the value corresponding to the requested key by using minimal network resources. At the physical layer, due to the broadcast nature of wireless channel, every transmission from a node can be heard by all its neighbors. At the network layer, the most widely used ad hoc routing protocols, such as AODV and DSR, are all based on broadcast. Whenever a source node wants to find a route to a destination node, the source node broadcasts a route request. If any node receiving the request is the destination node or knows a route to the destination, it sends a route reply back to the source; otherwise it will rebroadcast the route request. The complexity of a broadcast at the network and lower layers is $O(n)$. We know that the complexity of searching algorithms in the overlay is lower-bounded by $\log n$ achieved by DHT-based overlay searching algorithms. When we apply the overlay searching algorithm on the wireless mesh networks, it will need $O(\log(n))$ virtual hops in the overlay, while each virtual hop requires $O(n)$ physical hops to find the real route. So the complexity we can currently achieve is lower-bounded by $O(n\log n)$.

The proposed cross-layer searching algorithm achieves the complexity of $O(n)$ by taking advantage of the network layer broadcast to route the overlay searching request. When a node knows a key k and wants to find its corresponding value, it first maps k to a virtual network address *vAddress* using some known hash function. The hash function is employed to make *vAddress* unique and randomly distributed in the network address space. A magic mark may be embedded to *vAddress* so that other nodes can easily recognize that this is not a real network address. The source node then broadcasts a network route request for *vAddress*. Any node in the searching overlay will check received route request and compare the requested virtual address against all the keys it has. If the same key is found, the corresponding value will be sent back to the source node as a network route reply. Basically, this new algorithm combines the two searching processes in the overlay and the network layer into one single searching process. Some techniques can be further employed to enhance the algorithm. For example, when the value is sent back to the source node in the network route reply, an intermediate node relaying it can cache the (key, value) pair so that it can also respond when the same searching request comes next time. In addition to storing (key, value) pairs, a node can also save the corresponding *vAddress* for each pair and sort them by *vAddress*, so it can search the list and respond to a route request faster. Figure 5 details the cross-layer information exchanges between the overlay and the network layer in both the source and destination nodes.

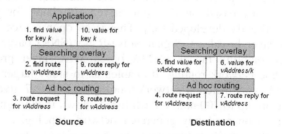

Fig. 5. Cross-layer searching algorithm

Figure 4c illustrates the advantage of using the proposed cross-layer searching algorithm. The physically shortest route can be quickly found. Experimentation and simulation results will be presented in the next section.

5 Results

Extensive experiments have been conducted on the implemented OverMesh platform. They demonstrate that the network-wide performance is affected by network topology, data traffic load, mobility, and even human activities. Due to the space limitation, we will only present the experimentation results related to the searching algorithm.

Figure 6 compares the cross-layer overlay searching (dot) and the OpenDHT overlay searching (circle) when 100 searching requests are sent. Two types of network topology are tested. In the mesh topology, a node can communicate with any of its neighbors, while in the linear topology, a node is forced to only communicate to one of its neighbors. It is evident that the cross-layer searching renders less response time. In average, the response time of cross-layer searching is 1.15 seconds, while the response time of OpenDHT searching is 3.55 seconds.

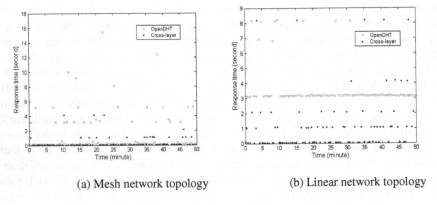

(a) Mesh network topology   (b) Linear network topology

Fig. 6. Experimentation results for searching

To compare the cross-layer searching and DHT searching in more scenarios, we further did simulations based on an event-driven simulator for mobile wireless networks that we recently developed [33]. This is an open source simulator which supports physical layer broadcast, Signal-Noise-Ratio (SNR) based packet capture, random way-point mobility model, CSMA based MAC layer protocol, and link quality based ad hoc network routing. It enables us to study larger mobile wireless mesh networks and many different network topologies. Figure 7 gives the simulation results when there are up to 100 nodes. Each result shown in the figure represents the average of 100 runs on randomly generated network topologies, with confidence interval of 95%. The cross-layer searching requires much less hops for each request than the DHT based overlay searching. This is mainly due to the mismatch between the overlay topology and the physical network topology. The larger number of physical hops in the DHT overlay searching introduces a lot more network routing control packets and therefore more packet collisions and losses. This will also make the response time much larger than that of the cross-layer searching.

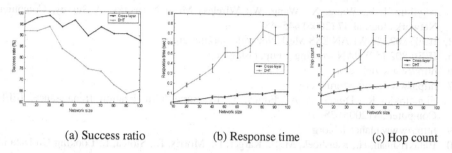

(a) Success ratio   (b) Response time   (c) Hop count

Fig. 7. Simulation result for searching with different network size

More simulations have been done when the number of concurrent searching requests is up to 10, when the maximum moving speed of each node increases from 1 m/s to 10 m/s, and when the packet loss zone increases from 0.1 db to 1 db. All these simulation results demonstrate that the cross-layer overlay searching significantly outperforms the DHT based overlay searching on the wireless mesh networks in terms of higher success ratio, less response time, and less physical hop count.

6 Conclusion and Future Work

This paper proposes a platform that realizes the conceptual architecture of overlays on wireless mesh networks, i.e. OverMesh. It provides an open and scalable testbed for developing and deploying distributed services and applications on wireless mesh networks. The current prototype is based on IEEE 802.11s pre-standard and private PlanetLab. A cross-layer overlay searching algorithm is proposed to enable fast searching on the resource limited wireless mesh networks.

We are currently pursuing several research directions based on the OverMesh architecture and the implemented platform. For example, extending the prototype to different scale wireless mesh networks or the multiple radio wireless networks; applying Xen virtual machine [26] and hardware virtualization technology [27]; investigating more cross-layer approaches to optimize resource usage and improve network efficiency in the wireless environment. We are also working on improving the current implementation. For example, customizing the boot CD in order to make the installation process easier; introducing Windows machines to the system; providing an open source toolkit so that other researchers can easily build their own system and conduct various research activities on wireless mesh networks.

References

1. Vicente, J., Rungta, S., Ding, G., Krishnaswamy, D., Chan, W., Miao, K.: OverMesh: Network Centric Computing. under submission to IEEE Communications Magazine (2006)
2. Chlamtac, I., Conti, M., Liu, J. J. -N.: Mobile Ad Hoc Networking: Imperatives and Challenges. Ad Hoc Networks 1 (2003) 13–64

3. Akyildiz, I. F., Wang, X., Wang, W.: Wireless Mesh Networks: A Survey. Computer Networks Journal 47 (2005) 445–487
4. IEEE 802.11s WLAN ESS Mesh Network working group
5. IEEE 802.15 WPAN working group, task group 5
6. http://www.zigbee.org
7. http://www.wimaxforum.org
8. Figueiredo, R., Dinda, P. A., Fortes, J.: Resource Virtualization Renaissance. IEEE Computer 38 (2005) 28–31
9. http://www.planet-lab.org
10. Balakrishnan, H., Kaashoek, M. F., Karger, D., Morris, R., Stoica, I.: Looking Up Data in P2P Systems. Communications of the ACM 46 (2003) 43–48
11. De Couto, D. S. J., Aguayo, D., Bicket, J., Morris, R.: A High-Throughput Path Metric for Multi-Hop Wireless Routing. ACM MobiCom (2003)
12. http://linux-vserver.org
13. http://ipw2200.sourceforge.net
14. Rhea, S., et al.: OpenDHT: A Public DHT Service and Its Uses. ACM SigComm (2005)
15. http://www.bamboo-dht.org
16. Shakkottai, S., Rappaport, T. S., Karlsson, P. C.: Cross-Layer Design for Wireless Networks. IEEE Communications Magazine 41(2003) 74–80
17. Cross-Layer Protocol Engineering for Wireless Mobile Networks. IEEE Communications Magazine 44 (2006)
18. Andersen, D. G., Balakrishnan, H., Kaashoek, M. F., Morris, R.: Resilient Overlay Networks. ACM SOSP (2001)
19. Eriksson, H.: Mbone: The Multicast Backbone. Communications of the ACM 37(1994) 54–60
20. Day, M., Cain, B., Tomlinson, B., Rzewski, P.: A Model for Content Internetworking. Internet RFC 3466
21. Xu, Z., Tang, C., Zhang, Z.: Building Topology-Aware Overlays using Global Soft-State. ICDCS (2003)
22. Nakao, A., Peterson, L., Bavier, A.: A Routing Underlay for Overlay Networks. ACM SigComm (2003)
23. http://www.vmware.com
24. http://denali.cs.washington.edu
25. http://user-mode-linux.sourceforge.net
26. http://www.xensource.com
27. Ublig, R., et al.: Intel Virtualization Technology. IEEE Computer 38(2005) 48–56
28. Sirer, E. G., Grimm, R., Bershad, B. N., Gregory, A. J., McDirmid, S.: Distributed Virtual Machines: A System Architecture for Network Computing. SIGOPS European Workshop (1998)
29. http://virtuoso.cs.northwestern.edu
30. http://www.emulab.net
31. Chandra, R., Bahl, P., Bahl, P.: MultiNet: Connecting to Multiple IEEE 802.11 Networks Using a Single Wireless Card. IEEE InfoCom (2004)
32. http://research.microsoft.com/netres/projects/virtualwifi
33. http://wireless-matlab.sourceforge.net
34. http://www.planet-lab.org/php/information.php?meeting_id=8
35. EverLab, http://www.cs.huji.ac.il/labs/danss/p2p/private-planetlab/index.html
36. OneLab, http://lsirwww.epfl.ch/PlanetLabEverywhere/slides/Lausanne.ppt

Adaptive Quality of Service Management for Next Generation Residential Gateways

Iván Vidal, Jaime García, Francisco Valera, Ignacio Soto, and Arturo Azcorra

Universidad Carlos III de Madrid
Avda. Universidad 30, 28911 Leganes - Madrid, Spain
{ividal, jgr, fvalera, isoto, azcorra}@it.uc3m.es

Abstract. The TISPAN workgroup inside ETSI is currently working on accommodating the IMS (IP Multimedia Subsystem) architecture, which has been created for the mobile world, to the fixed scenario where there is a new important element to be taken into account: the Residential Gateway (RGW). This element is typically considered as a customer device where providers do not usually have anything to configure. However, in order to achieve real end-to-end Quality of Service (QoS) this cannot be true anymore.

This paper focuses on the way that a RGW is capable of configuring itself (an interface with the providers is also available), regarding Quality of Service parameters, into a Next Generation Network (NGN) scenario. The proposed RGW architecture is also flexible enough so as to adapt the QoS management mechanism to different possible scenarios, e.g. configured by the provider, by the customer or even autoconfigured by the RGW itself. A specific scenario, where a RGW is deployed in the TISPAN NGN architecture, will be explained and validated to proof the concept of the RGW architecture.

1 Introduction

Nowadays a lot of work around Next Generation Networks (NGN) has been done by the scientific community to achieve a real integration of every access technology actually deployed to provide triple-play services (video, data and voice). This work is still unfinished due to the high complexity of the problem and there are in fact many initiatives in progress. ETSI TISPAN is one of them and its philosophy is to try to conjugate different standards together and improve them to be fully compatible with the NGN architecture. The TISPAN defined what a NGN is in [1]:

"A Next Generation Network is a packet-based network able to provide services including Telecommunication Services and able to make use of multiple broadband, QoS-enabled transport technologies and in which service related functions are independent from underlying transport related technologies. It offers unrestricted access by users to different service providers. It supports generalised mobility which will allow consistent and ubiquitous provision of services to users."

A. Helmy et al. (Eds.): MMNS 2006, LNCS 4267, pp. 183–194, 2006.

As a result of the ongoing work, the first release for TISPAN Next Generation Network (NGN) [2] was published at the beginning of 2006. The core network of TISPAN NGN is based upon the IMS, as defined in 3GPP Release 6 and 3GPP2 revision A for IP-based multimedia applications (although IMS is conceptually designed to be independent from the technology used in the access network, the standards developed by the 3GPP from release 5 are mainly focused on the UMTS IP connectivity access network). Due to this decision, TISPAN architecture obtains many advantages from merging both wireless and wired worlds because the only need fact is an IMS adaptation to the fixed world. But this work is not an easy task and a lot of discussions are still open. One of these discussions is related with the RGW configuration, because in the mobile world this entity does not exist. To date, there is no way to remotely configure the RGW QoS parameters by a NGN operator so there is still a gap in the architecture. Actually, the RGW is considered by the TISPAN as a customer equipment with no logical interfaces towards the network so as to be able to configure it. In this paper a flexible RGW architecture is proposed allowing the own gateway to autoconfigure itself using the signalling interchanged by the customer devices and the network. This architecture extends and demonstrate the concepts published in a previous work [3] where just the basic functionality was described, in order to explain the work developed in the MUSE European Project [4] where the main goal is to research the European next generation network. In MUSE the RGW is considered a key component and an entire Task Force is focused in its study and definition.

The rest of this article is organised as follows. The TISPAN NGN QoS management architecture is reviewed in next Sect. 2. The complete RGW architecture is proposed in Sect. 3 while the particular part where the QoS autoconfiguration is extended is analysed in Sect. 4. Section 5 concludes the article with the main contributions of this work.

2   TISPAN NGN QoS Management Overview

In this section, just a brief overview of TISPAN NGN QoS management is presented. The functional architecture of TISPAN NGN in release 1 is described in detail in [5] and the rest of the article follows the terminology defined by TISPAN. Figure 1 covers a simplified overview of this architecture to facilitate the reading of this document.

2.1   Resource and Admission Control in TISPAN NGN

The Resource and Adsmission Control Subsystem (**RACS**) is the TISPAN NGN subsystem that provides QoS reservation mechanisms over the transport layer. In this way, RACS provides the ASFs and the Service Control Subsystems with means to request and reserve resources from the transport networks which are under its control.

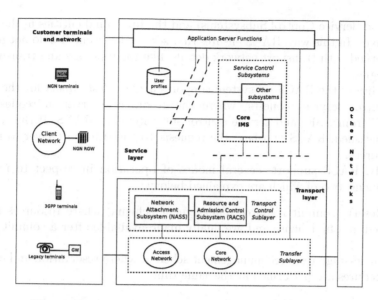

Fig. 1. Functional architecture of TISPAN NGN, release 1

On the other hand, in the NGN scope, the access network is viewed as the most critical segment to provide end-to-end QoS. For this reason, Release 1 of TISPAN NGN is mainly focused on this segment of the transport network in respect to QoS provision, assuming that QoS in the Core Network is provided by other means.

Therefore, RACS scope in TISPAN NGN Release 1 is limited to the access network, to the interconnection point between the access network and the Core Network and to the interconnection point between Core Networks. The release does not require the RACS subsystem to cover the Core Network itself or any equipment located in client premises.

The basic RACS functionalities in TISPAN NGN are indicated below:

- **Policy Control.** RACS applies to resource reservation requests a set of policy rules to check if these requests can be authorised and to determine how must they be served. Policy control is also performed in the access network, applying network policies specific to each particular access line.
- **Admission Control.** RACS verifies the authorisation to reserve resources on the access network based on user profiles, the access network policies and the resource availability.
- **Resource reservation.** RACS provides the means to reserve bearer resources on the access network.
- **NAT/Gate Control.** RACS controls NAT and NAT-T functionalities and performs gate control functions, at the limit between the access and the core networks and in the limit between core networks.

So, RACS provides the Service Layer with a single interface to request transport control services, acting as an intermediary between the service layer entities

(ASFs and Service Control Subsystems) and the functional entities in the transfer sublayer. In this way, RACS ensures that service layer entities do not need to be concerned with transport network details, like the topology and transmission technologies.

In addition, the RACS subsystem supports a *QoS Push Model* for the transport control service requests, where the resource reservation requests are "pushed" from ASFs and service control subsystems to RACS. If the requests can be satisfied, RACS "pushes" the requests to the transport layer to reserve the resources.

Finally, RACS supports several modes of operation in respect to resource management, two of which are explained below:

– A Reserve-commit resource management schema, where resources are reserved in a first phase and are finally made available after a commit procedure.
– A single-stage resource management schema, where reservation and commit procedures are performed at the same time.

QoS Control Models. The RACS subsystem supports two different models for QoS control over the transport network. These models are the following:

– **Guaranteed QoS.** In this model the QoS is guaranteed with absolute bounds on some or all of its parameters, like throughput or jitter. Guaranteed QoS is configured on the access network with the application of techniques such as throughput control and traffic policing in the IP edge node. These techniques may also be applied to the Access Node or to the equipments in the client premises.
– **Relative QoS.** In this model the QoS is provided by class based differentiation. This QoS differentiation is configured in the IP edge node of the access network, where functionalities like packet marking are provided. RACS should be aware that some equipment in the client premises may also provide QoS differentiation, for example through packet marking. This marking should only have effect if it is required by the operator.

RACS Functional Architecture. Figure 2 shows the functional architecture of the RACS subsystem.

The Application Function (**AF**) interacts with the RACS subsystem to request transport-layer control services for QoS provisioning to services. This function is implemented in some functional entities from the service layer, such as the ASF and the P-CSCF of the Core IMS. The AF converts QoS information from the application layer to QoS information which is suitable for the RACS subsystem, and includes this information in a request message, which is sent to the SPDF through the Gq' interface. The details of the protocol used in the interface Gq' are specified in [6].

The Service Policy Decision Function (**SPDF**) authorises the request, checking the information contained on it against the local policy established for the

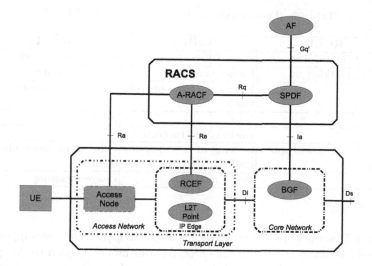

Fig. 2. Functional architecture of RACS

requester AF. If the request is successfully authorised, the SPDF determines if it must contact the A-RACF and/or the BGF to provide the transport-layer control service.

The Access-Resource and Admission Control Function (**A-RACF**) is always in the access network, and provides the functionality for admission control and resource reservation over the access network to the SPDF. The A-RACF can accept or reject the requests received from the SPDF based on the admission control mechanisms.

The Resource Control Enforcement Function (**RCEF**) is located in the IP edge node of the access network, and provides the RACS with the means to apply the traffic policies that guarantee the resource reservation. The Border Gateway Function (**BGF**) provides the interface between two IP domains. Release 1 of TISPAN NGN identifies two types of BGF: the Core BGF (**C-BGF**), which is located between the access and the core network (in the Core Network side), and the Interconnection BGF (**I-BGF**), which is located in the limit between two Core Networks.

Table 1 offers an overview of the functionalities provided by the RCEF, C-BGF and I-BGF, as they are explained in [7].

2.2 Session Establishment and Resource Reservation

The session establishment process in IMS is based on the Session Initiation Protocol (SIP), and involves an end-to-end signalling dialogue between the terminals participating in the session. To negotiate the parameters associated with the media which is going to be transferred during the session, such us the type of media streams, codecs or IP addresses and ports, the Offer/Answer model of

Table 1. Functionality of RCEF, C-BGF and I-BGF

RCEF	C-BGF	I-BGF
Open/Close Gates	Open/Close Gates	Open/Close Gates
Packet marking	Packet marking	Packet marking
	Resource allocation (per flow)	Resource allocation (per flow)
	NAT	NAT
	Hosted Nat traversal	
Policing of down/uplink traffic	Policing of down/uplink traffic	Policing of down/uplink traffic
	Usage metering	Usage metering

the Session Description Protocol (SDP) is used. SDP provides the support to describe multimedia sessions, and the Offer/Answer model applied to this protocol allows the end terminals to reach an agreement about the session description. The complete specification of SIP can be found in [8], whereas the Offer/Answer model with SDP is detailed in [9].

In the IMS architecture, the P-CSCF (Proxy-Call Session Control Function) is the functional entity which acts as the entry point for the users to the system. So, all the SIP signalling messages that come from or go to the user terminal must necessarily pass through this functional entity.

In TISPAN NGN Release 1, the P-CSCF in the Core IMS implements the AF functionality indicated in section 2.1, interacting with RACS to request QoS provision for the services negotiated between the end users. As it is detailed in [6], the P-CSCF sends service information to the RACS subsystem (i.e. to the SPDF) after receiving every SIP message with a SDP answer payload. This service information is derived from that SDP answer and from its corresponding SDP offer. Jointly, the SDP *offer* and SDP *answer* contain enough information to describe the session as it has been negotiated up to that moment, such us the IP addresses, ports and bandwidth requirements for the IP flows that will be transferred. Annex B in document [6] describe the mapping process that must be performed to convert SDP information to relevant service information to be transferred to the RACS subsystem.

As an example, consider the scenario proposed in Fig. 3 where it is assumed that two users want to establish a VoIP call through the TISPAN NGN, being both of them connected to the NGN through different xDSL access networks, and that only public IP addressing schemes are in use. The figure only shows the signalling and the QoS reservation process from the point of view of the user originating the call. The example assumes a reserve-commit resource management schema.

The procedure is as follows:

1-3. The terminal sends a SIP INVITE request, including a SDP offer, to the P-CSCF. The P-CSCF processes the request and forwards it to the

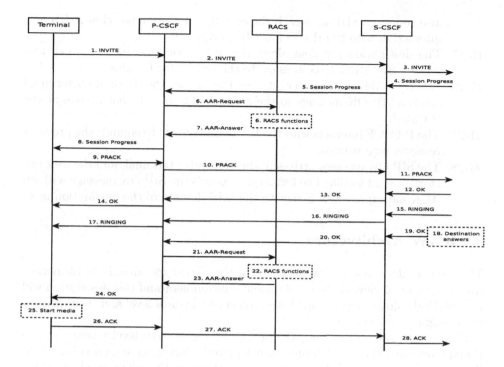

Fig. 3. Example of IMS signalling and QoS reservation

proper S-CSCF. After that, the S-CSCF processes the request and forwards it to the IMS domain of the destination user.

4-5. The destination terminal answers back with a Session Progress message, which is received in the S-CSCF. The S-CSCF processes it and forwards it to the P-CSCF.

6-8. The P-CSCF derives the service information from the SDP offer and answer, and then sends a resource reservation request to the RACS subsystem. RACS authorises the request and performs the corresponding admission control procedures, reserving the necessary resources from the transport network. Finally, it confirms to the P-CSCF that the resource reservation procedure has succeed. Finally, the P-CSCF forwards the Session Progress message to the terminal.

9-11. The terminal decides the final parameters for the session and confirms the reception of the Session Progress with a PRACK message. This message may also contain a SDP payload, that may be the same as the one that was received in the previous message or a subset. The terminal can make a new SDP offer in this message, or later using SIP UPDATE messages. Anyway, each SDP offer/answer pair will cause a new interaction with RACS from the P-CSCF.

12-14. The destination terminal acknowledges the PRACK with an OK message and if the PRACK message included an SDP offer, the OK message will also contain a SDP answer. If the session description has changed, a new interaction with the RACS must be performed.

15-17. The destination terminal alerts the user about the incoming call and sends a SIP RINGING message to the originator terminal.

18-20. When the destination user answers the phone, the destination terminal sends a SIP OK message to the originator terminal, that arrives at the P-CSCF.

21-23. The P-CSCF interacts with the RACS subsystem to commit the previous resource reservations.

24-28. The SIP OK message arrives to the originator terminal, and the user can start to send media. The terminal responds the SIP OK message with an ACK message, which is sent to the IMS domain of the destination user.

3   RGW Architecture

This section describes the architecture of the prototype already implemented from the point of view of the development environment and this description will be classified using a bottom-up view starting at the data level and then defining the configuration process.

Figure 4 represents the complete picture at the bottom level where all functional blocks and their relationships are depicted[1]. Incoming and outgoing traffic flows are represented and the two separate paths show that these two flows never use the same resources at Click! level. Dotted arrows represent unknown outgoing traffic. Click! level sends these packets to the CSD (Click! Signalling Dispatcher) to treat them and then it sends the packets to the corresponding Signalling Process (SP) to handle it. Finally, the SP returns the packets to Click! level. Dashed arrows are frame copies that Click! sends to the CSD or the IMS due to special characteristics (signalling frames, for example).

These are the functional blocks for this level:

- **User Classifier:** The function of this block is to recognise flows depending on the user (administrator) preferences. The user can add, reorder or remove flow definitions and this change will modify the functionality of this block. In the downstream direction, the user can select that a particular flow can be replicated to all in-home interfaces selecting the multicast option.
- **Dispatcher:** Based on the p-bits field, this module introduces a packet in one of the four possible queues.
- **Queues:** In Click!, the implementation of these queues is based on the invocation of four different queue elements. Each queue represents a different CoS. There are several ways to accomplish the requirements imposed by a specific CoS. For example, a fix size queue can be used to avoid queue delays.
- **Scheduling:** Working with two or more queues implies the use of some algorithm to extract a packet from one queue at each time. It is even more complex to select the right one when priority queues exist. There are many

---

[1] The core of this prototype is implemented using the Click! modular router [10].

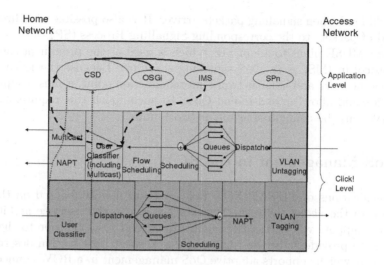

Fig. 4. RGW Functional Blocks

scheduling algorithms to treat this problem: Priority Queuing, Weighted Fair Queuing, Class-based Weighted Fair Queuing, etc. and a Click! element called PrioSched that implements a Priority-like Queuing is used.
- **Policing:** The aim of this functional block is to limit the flows rate for every CoS. When an excess rate is detected, this block can either discard packets or change their CoS tag.
- **VLAN untagging:** Removes the 802.1Q/p tags from the packets.
- **Flow scheduling:** Detects "important" packets as signal packets and send copies to the CSD. It is useful to detect new flows and their characteristics (CoS for example).
- **NAPT:** For the upstream traffic it detects new sessions and then creates new entries in the NAPT table to change the original source port with the corresponding identification. For the downstream direction, this box just has to change the destination port and IP address with the matching entry from the NAPT table. Upstream and downstream directions are not different boxes because they must share the NAPT table.
- **VLAN tagging:** Depending on the user/administrator preferences, outgoing packets will be marked with 802.1Q/p tags and bits to add a priority to the frames. This information (the corresponding p-bit assignment) is pre-configured in the first stage and configured by the network in the final prototype. When a flow is not configured, this block forwards the packet to the CSD. The CSD can then discard the packet or reconfigure the VLAN tagging functional block (to recognise this new flow) and inject the packet again.
- **Multicast:** This block copies all incoming frames to all in-home interfaces (like most Ethernet switches do).
- **CSD:** This software has been developed in Java to allow an easy and quick portability to other platforms. The CSD will configure and reconfigure the

Click! level when signalling packets arrives. It is also possible to redirect this kind of packets to the corresponding Signalling Process (SP).

- **IMS:** MUSE plans to adopt SIP, which is used as the prominent signalling protocol in IMS, as its QoS signalling standard. Because the RGW is the access network and home network interconnection point, it must couple both worlds and allow an end-to-end QoS. To accomplish this requirement, the RGW must be configured as any network node.

4   QoS Management in the RGW

The specifications of TISPAN NGN Release 1 are mainly focused on the QoS provision in the access network. Nevertheless, a complete service architecture with QoS support will necessarily require to extend to QoS scope to the client premises, to provide end-to-end QoS. In this respect, this section describes an architecture which supports adaptive QoS management in a RGW connected to a NGN compliant with TISPAN NGN Release 1. The proposed architecture for the RGW, from the point of view of QoS management, is shown in Fig. 5.

The **NGN SP** is the Signalling Process that will manage, in the RGW, the SIP signalling used to negotiate end-to-end services over the TISPAN NGN. It will process all the SIP signalling messages interchanged between the NGG core IMS and the customer equipment. The NGN SP will provide the following functionalities:

- **P-CSCF functionality:** After receiving every SIP message containing a SDP answer, it will derive the corresponding service information from the SDP offer and the SDP answer as it is specified in annex B of [6]. This information will be provided to the A-RACF functionality of the NGN SP. On the other hand, after receiving a SIP OK message corresponding to a SIP INVITE message, it will contact the A-RACF functionality of the NGN SP to start the commitment process for the reserved resources.
- **A-RACF functionality:** With the service information provided by the P-CSCF functionality, the A-RACF functionality will perform admission control functions, verifying if the QoS demands can be satisfied in the RGW and the home network with the available resources. The A-RACF functionality will support a reserve-commit resource management schema and the QoS control models defined for the RACS in [7], i.e. guaranteed QoS and relative QoS.
- **SIP signalling proxy:** This functionality is necessary in the Signalling Proxy Scenario (SPS) proposed in [3]. In this scenario the NGN SP would behave as a signalling proxy on behalf of legacy terminals, by generating the SIP signalling associated with the upstream and the downstream traffic.

The **RCEF** will implement the functionality of a Resource Control Enforcement Function as it is detailed in [5] and [7], i.e. opening/closing gates, packet marking and policing of down/uplink traffic. The RCEF will apply the traffic

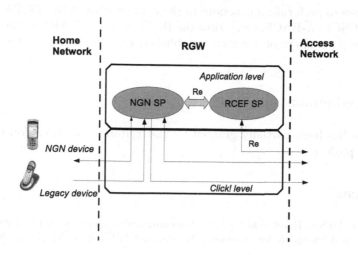

Fig. 5. RGW QoS management architecture

policies established by the A-RACF functionality of the NGN SP, guaranteeing the reservation of resources in the RGW.

Finally, the RCEF will provide a interface fully compatible with the Re interface defined in the RACS architecture detailed in [7]. This way, it will be possible a future scenario where the RACS subsystem in the NGN access network may directly access the RCEF function in the RGW to manage the QoS configuration in the client premises.

5 Conclusions

To date, there are a lot of telecommunications infrastructures capable of carrying different types of information: a specific one for the fixed telephone, a different technology for the mobile phones (moreover GSM, GPRS, UMTS, etc. are totally different) and other equipments to transport data (the Internet network). A telecommunications operator providing voice, video and data services needs to bring together all these technologies in a more efficient one to reduce costs. The IP protocol could be used to bring Triple Play services but many problems have to be solved first and the QoS is one of them. ETSI TISPAN Release 1 defines how to achieve QoS deliveries up to the access network but, since the RGW is not considered part of the access network by TISPAN, the end user would not perceive that QoS when its own home network is a bottleneck.

In this article we presented a flexible RGW architecture where both, the RGW resources and the home network available bandwidth are taken into account to accept or not a given connection. The schema proposed is valid for a RGW aware NGN architecture, where the access and reservation systems configure the RGW and in a RGW unaware one. For the later, the RGW intercepts SIP requests and responses to perform some actions in the same way as in the TISPAN NGN devices (P-CSCF, A-RACF, etc.). How the RGW manages NAPT scenarios and updates the parameters of a certain established connection are let for further study.

Acknowledgements

This article has been partially granted by the European Commission through the MUSE project.

References

1. TISPAN: ETSI TR 180 000 V1.1.1: "Telecommunications and Internet converged Services and Protocols for Advanced Networking (TISPAN); NGN Terminology." (2006)
2. TISPAN: ETSI TR 180 001 V1.1.1: "Telecommunications and Internet converged Services and Protocols for Advanced Networking (TISPAN); NGN Release 1; Release definition." (2006)
3. Guerrero, C., Garcia-Reinoso, J., Valera, F., Azcorra, A.: Qos management in fixed broadband residential gateways. LNCS **3754** (2005) 338–349 8th International Conference on Management of Multimedia Networks and Services (MMNS 2005).
4. MUSE: Multi Service Access Everywhere. Internet (2006) http://www.ist-muse.org/.
5. TISPAN: ETSI ES 282 001 V1.1.1: "Telecommunications and Internet converged Services and Protocols for Advanced Networking (TISPAN); NGN Functional Architecture Release 1" (2005)
6. TISPAN: ETSI TS 183 017 V1.1.1: "Telecommunications and Internet converged Services and Protocols for Advanced Networking (TISPAN); Resource and Admission Control; DIAMETER protocol for session based policy set-up information exchange between the Application Function (AF) and the Service Policy Decision Function (SPDF): Protocol specification." (2006)
7. TISPAN: ETSI ES 282 003 V1.1.1: "Telecommunications and Internet converged Services and Protocols for Advanced Networking (TISPAN); Resource and Admission Control Sub-system (RACS); Functional Architecture." (2006)
8. Rosenberg, J., Schulzrinne, H., Camarillo, G., Johnston, A., Peterson, J., Sparks, R., Handley, M., Schooler, E.: SIP: Session Initiation Protocol. RFC 3261 (Proposed Standard) (2002) Updated by RFCs 3265, 3853, 4320.
9. Rosenberg, J., Schulzrinne, H.: An Offer/Answer Model with Session Description Protocol (SDP). RFC 3264 (Proposed Standard) (2002)
10. Kohler, E., Morris, R., Chen, B., Jannotti, J., Kaashoek, M.F.: The Click Modular Router Project. Internet (2006) http://www.read.cs.ucla.edu/click/.

On the Location-Awareness of Bandwidth Allocation and Admission Control for the Support of Real-Time Traffic in Class-Based IP Networks[*]

Stylianos Georgoulas[1], Panos Trimintzios[1,2], George Pavlou[1], and Kin-Hon Ho[1]

[1] Centre for Communication Systems Research, University of Surrey
Guildford, Surrey, GU2 7XH, United Kingdom
[2] ENISA, PO Box 1309, GR-71001, Heraklion, Crete, Greece

Abstract. The support of real-time traffic in class-based IP networks requires reservation of resources, accompanied by admission control in order to guarantee that newly admitted real-time traffic flows do not cause any violation to the Quality of Service (QoS) perceived by the already established ones. In this paper we highlight certain issues with respect to bandwidth allocation and admission control for supporting real-time traffic in class-based IP networks. We investigate the implications of topological placement of both bandwidth allocation and admission control schemes. We show that their performance depends highly on the location of the employed procedures with respect to the end-users and the various network boundaries. We conclude that the strategies for applying these schemes should be location-aware, because their performance at different points in a class-based IP network can be different and can deviate from the expected performance. Through simulations we also provide a quantitative view of these deviations.

1 Introduction

Class-based service models, such as the Differentiated services (Diffserv) model, offer a scalable approach towards QoS. This is achieved by grouping traffic with similar QoS requirements into one of the engineered traffic classes and forwarding it in an aggregate fashion. By allowing traffic aggregation, networks that deploy class-based service models can take advantage of statistical multiplexing, which allows for efficient use of the resources. In order to provide QoS guarantees, a network supporting different classes must also deploy admission control to control the amount of traffic injected into the network so as to prevent overloads that can lead to QoS violations.

In this work we focus on the support of *real-time* traffic in an IP network with class-based service support. We assume that in such a network, there exists end-to-end isolation between the UDP real-time traffic and the TCP controlled data traffic. This is essential for guaranteeing QoS [1]. We can achieve this isolation by using different queues for the two types of traffic. An example of this scenario could be a Diffserv network where the UDP real-time traffic is classified to use a higher priority Assured Forwarding (AF) Per-Hop Behavior (PHB), whereas the TCP controlled data traffic is classified to use a lower priority, possibly best-effort, forwarding PHB.

---

[*] This work was undertaken in the context of the FP6 Information Society Technologies AGAVE (IST-027609) project, which is partially funded by the Commission of the European Union.

A. Helmy et al. (Eds.): MMNS 2006, LNCS 4267, pp. 195–207, 2006.

We define as *real-time* traffic flows, flows that have low delay and jitter require-ments, and a bounded packet loss rate (PLR) requirement and we focus on the PLR requirement. For services, such as Voice or Video, a certain amount of packet loss can be acceptable [2] without significant quality degradation. Therefore, such services do not need the 'virtual wire' (Expedited Forwarding (EF) in the Diffserv model) treat-ment [3]. The low delay and jitter requirements are likely to be met in a high-speed network [2] and offline traffic engineering actions can be additionally taken so that delay and jitter are kept within low bounds. For example, the delay requirement can be taken into account at the network provisioning phase by a) configuring small queues for the real-time traffic so as to keep the per-hop delay small, and b) control-ling the routing process to choose paths with a constrained number of hops. Jitter can remain controlled as long as the flows are shaped to their peak rate at the network in-gress [4]. Also, the deployment of non-work conserving scheduling in routers for the real-time traffic class can be beneficial for controlling jitter [5]. Therefore, we assume that the real-time traffic flows can be shaped to their peak rate at the network ingress and that the scheduling mechanism for the real-time traffic class is priority scheduling with a strict bandwidth limit and with First-In-First-Out (FIFO) service discipline.

In this work we aim to demonstrate how topological placement, that is the location of the employed bandwidth allocation and admission control schemes with respect to the end-users and the various network boundaries, can affect their performance. Be-fore proceeding, we will briefly describe the bandwidth allocation schemes suitable for aggregating real-time traffic and also describe the admission control schemes that can be used for real-time traffic in class-based IP networks, so that later we can refer to them and point out how the results of our work relate to them.

2 Bandwidth Allocation Schemes

Bandwidth allocation schemes can be divided in two main categories. The first cate-gory comprises schemes based on *bufferless statistical multiplexing*. Bufferless statis-tical multiplexing aims to ensure that the combined arrival rate of the multiplexed sources exceeds the allocated capacity only with very small probability. Examples of such bandwidth allocation schemes can be found in [6, 7]. The second category com-prises schemes based on *buffered statistical multiplexing*. Contrary to bufferless mul-tiplexing, buffered multiplexing allows an input rate excess, with surplus traffic being stored in large buffers. Examples of such bandwidth allocation schemes can be found in [6, 7, 8]. Generally speaking, both categories take into account factors such as the number and characteristics of flows, the required loss rate and, in case of buffered multiplexing, the available buffer size and derive the required capacity (*effective bandwidth*) needed for the loss rate to be kept below the required threshold.

Each of the two categories has its merits but also its drawbacks. The dynamics leading to an overload event in a bufferless system are much simpler than those of a buffered system [9]. On the other hand, buffered multiplexing allows for higher utili-zation for the same loss rate [9]. We need to stress though, that bufferless multiplex-ing is just a model abstraction [9]. For packetized traffic, a small buffer for packet scale queuing is needed to account for simultaneous packet arrivals from distinct flows [8]. However, in this case the buffer is not used for storing significant amounts of excess traffic and is, therefore, not involved in bandwidth estimations.

3 Admission Control Schemes

We can broadly divide the various admission control schemes found in the literature into three categories: endpoint admission control (EAC), traffic descriptor-based admission control (TDAC), and measurement-based admission control (MBAC).

EAC is based on some metrics applied to probing packets sent along the transmission path before the flow is established [10]. A requirement is for the end-to-end route to be the same for probing packets and flows. For reasonably bounded setup delays the metrics do not depict stationary network states but snapshots of network status, which can result to a quite unrealistic picture of the network congestion and, furthermore, simultaneous probing by many sources can lead to a situation known as thrashing [10]. That is, even though the number of admitted flows is small, the cumulative level of probing packets prevents further admissions.

TDAC is based on the assumption that traffic descriptors are provided for each flow prior to its establishment. This approach achieves high utilization when the traffic descriptors used by the scheme are appropriate. Nevertheless, in practice, it suffers from several problems [11]. One is the inability to come up with appropriate traffic descriptors before establishing the flow. Another problem is that the traffic descriptors and the associated QoS guarantee define a contract between the flow and the network. Therefore, the need to police based on this traffic specification arises, which is difficult for statistical traffic descriptors [11]. Deterministic models, such as token buckets, are easy to police but they fail to provide a sufficient characterization to extract a large fraction of the potential statistical multiplexing gain [11].

MBAC tries to avoid the problems of the other approaches by shifting the task of traffic characterization to the network [11]. That means that the network attempts to "learn" the characteristics of existing flows through real-time measurements. This approach has a number of advantages. First, the specified traffic descriptors can be very simple, e.g. peak rate, which is easy to police. Second, a conservative specification does not result in over-allocation of resources for the entire duration of the service session. Third, when traffic from different flows is multiplexed, the QoS experienced depends on their aggregate behavior, the statistics of which are easier to estimate than those of an individual flow. However, relying on measured quantities raises certain issues, such as estimation errors, flow-level dynamics and memory related issues [11].

4 The Effects of Topological Placement

In this section we will show how the topological placement of the functionalities can affect the performance of bandwidth allocation and admission control schemes.

For the needs of our work we will initially adopt the normal distribution based bufferless statistical multiplexing approach. According to [6], when the effect of statistical multiplexing is significant, the distribution of the stationary bit rate can be accurately approximated by a Gaussian distribution. In [12] it is suggested that the aggregation of even a fairly small number of traffic streams is usually sufficient for the Gaussian characterization of the input process.

In this case, the effective bandwidth of N multiplexed sources is given by [6, 7]:

$$C \simeq m + a'\sigma \text{ with } a' = \sqrt{-2\ln(\varepsilon) - \ln(2\pi)} \tag{1}$$

where $m = \sum_{i=1}^{N} m_i$ is the mean aggregate bit rate, $\sigma = \sqrt{\sum_{i=1}^{N} \sigma_i^2}$ is the standard deviation of the aggregate bit rate, and ε the upper bound on allowed loss rate. We will denote the function of eq. 1 as $eff(S, PLR)$, where S is the set of aggregated sources and PLR is the PLR value used in the calculation of the effective bandwidth C.

We will present our study with a list of scenarios using a two-level tree topology, which allows us to illustrate the mains points of our study, while being simple enough to suit the nature and computational demands of the required packet-level simulations.

4.1 Scenario I: The Effects of Aggregate Bandwidth Allocation

Initially we consider the scenario depicted in Fig. 1.

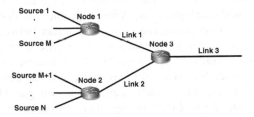

Fig. 1. Topology for assessing the effects of topological placement

In this scenario it is assumed that a set of sources, S_i, $i = 1,...,M$, are aggregated at node 1 and that another set of sources, S_i, $i = M + 1,...,N$, are aggregated at node 2. We assume that the sources connect to nodes 1 and 2 with direct links with negligible congestion, and that all of them will be eventually aggregated in the same traffic class at link 3. The capacity reserved in link 1 for the first set of sources is:

$$C_1 = eff(\{S_1,...S_M\}, PLR_1) \tag{2}$$

where PLR_1 is the packet loss rate budget for the real-time traffic class aggregate in link 1. Similarly, for the second set of sources, the capacity reserved in link 2 is:

$$C_2 = eff(\{S_{M+1},...S_N\}, PLR_2) \tag{3}$$

where PLR_2 is the allowed packet loss rate budget for the real-time traffic class aggregate in link 2. Since all the sources will be aggregated using the same class, the required bandwidth to be allocated in link 3 for their aggregation is given by:

$$C_3 = eff(\{S_1,...S_N\}, PLR_3) \tag{4}$$

where PLR_3 is the allowed packet loss rate budget for the real-time traffic in link 3.

This scenario could correspond to a situation where end-users (the $1,...,N$ sources) connect to the edge routers (nodes 1 and 2), which then connect to the metro/backbone router (node 3) through access links 1 and 2.

As it can be easily proven [13], packet loss rate parameters are multiplicative. That means that for a set of sources that traverse a sequence of links, l_i, $i = 1,...,L$, with packet loss rates PLR_i, the total packet loss rate PLR_{total} can be approximated by:

$$PLR_{total} = 1 - \prod_{i=1}^{L}(1 - PLR_i) \tag{5}$$

which, in turn, becomes additive for low values of PLR_i :

$$PLR_{total} = \sum_{i=1}^{L} PLR_i \tag{6}$$

Assuming that $PLR_1 = PLR_2 = PLR_3$, that is the allocated capacities at links 1, 2 and 3 for the real-time traffic class are such that allow for the same packet loss rate budget at all links, the expected overall upper bound on total, end-to-end in our topology, packet loss rate for the aggregate sources should be:

$$PLR_{total} = PLR_1 + PLR_3 = PLR_2 + PLR_3 \tag{7}$$

Our study initially aims to examine whether the actual total packet loss rate is bounded by the above expression. In order to do so, we run simulations using the network simulator *ns-2* [14]. For the simulations we use two example values for the target link packet loss rates, 0.01 and 0.001. We assume, without loss of generality, that the same number of sources is aggregated in both links 1 and 2, i.e. $M = N/2$. This means that the total packet loss rate, end-to-end in our topology, should not exceed 0.02 and 0.002 respectively. We also consider the case where the capacity in link 3 is provisioned so that $PLR_3 = 0$. This happens when $C_3 = C_1 + C_2$, which means that only links 1 and 2 incur losses and in link 3 the real-time traffic aggregates from nodes 1 and 2 are treated using peak rate allocations.

We consider three scenarios for the N traffic sources: a) all sources are *VoIP* sources with peak rate 64kbps and exponentially distributed ON and OFF periods with average durations 1.004sec and 1.587sec respectively (mean rate 24.8kbps, standard deviation of rate 31.18kbps) [15], b) all sources are *Videoconference* sources with mean rate 3.89Mbps, peak rate 10.585Mbps and standard deviation of rate 1.725Mbps [16], and c) that we have a *mixture* of both VoIP and Videoconference sources. We fix the packet size to 100bytes (constant packet size seems to be a reasonable assumption for Voice and Video communications [17]) and since the real-time traffic class is assumed to be isolated from other classes, we do not consider any best-effort or any other traffic classes and we simulate the real-time traffic class as being serviced by queues running at the speed of their bandwidth limit.

In the following figures, PLRa corresponds to the (average) packet loss incurred at links 1 and 2, while PLRb corresponds to the total packet loss for the cases where $PLR_1 = PLR_2 = PLR_3$ and they are given as a function of the mean aggregate bit rate of all sources S_i, $i = 1,...,N$ (x-axis).

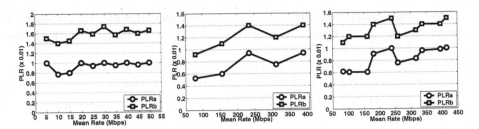

Fig. 2. Incurred PLR for *VoIP* (left), *Videoconference* (center) and *mixed* (right) traffic sources for target link PLR 0.01

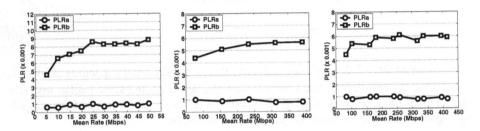

Fig. 3. Incurred PLR for *VoIP* (left), *Videoconference* (center) and *mixed* (right) traffic sources for target link PLR 0.001

From the above figures we can see that in all cases, the target packet loss rate in links 1 and 2 (PLRa) is satisfied. Regarding the total packet loss rate, when the bandwidth in link 3 is set so that link 3 also incurs losses (PLRb) we can see that it is kept below the target total packet loss rate 0.02 (see Fig. 2), but not in the case where the target total packet loss rate is set to 0.002 (see Fig. 3). Since the target packet loss rate in links 1 and 2 is always satisfied, that means that in the latter case, link 3 incurs losses that are much higher than the target packet loss rate budget at that link.

These results suggest that even though the original traffic descriptors are valid at the first points of aggregation, they may not be valid in transit nodes such as in node 3. This is because sources become correlated and their characteristics are altered as they traverse links and multiplexers. Therefore, using the original traffic descriptors in transit nodes can lead to erroneous bandwidth allocation decisions. This traffic profile deformation has been verified in the past [8, 18, 19] and analytical models for evaluating it for specific types of individual sources and under specific network conditions and assumptions have been presented [8].

In this work, we try to go one step further and quantify the effects that this aggregate traffic profile deformation can have on the incurred packet loss and, therefore, on the perceived QoS when short-range dependent (VoIP), long-range dependent (Videoconference), as well as a mixture of short-range and long-range dependent real-time traffic sources are aggregated in the same traffic class. In such general cases, obtaining a closed form solution to quantify its effects can become very difficult.

4.2 Scenario II: Quantifying the Effects of Traffic Profile Deformation

In order to quantify the effects of traffic profile deformation further downstream from the first hop node, we proceed as follows. We use the same simulated topology, traffic volume and types of sources, as in the previous scenario. We set the capacities allocated to the real-time traffic class in links 1 and 2 equal to the sum of peak rates of the sources aggregated in links 1 and 2 (links 1 and 2 are transparent to the sources with respect to packet loss). For link 3 we distinguish two cases. In the first case, we merge the two demands –one composed of the sources S_i, $i = 1, ..., M$ and the other composed of the sources S_i, $i = M + 1, ..., N$ – in one bandwidth allocation in link 3:

$$C_3 = eff(\{S_1, ... S_N\}, PLR_3) \tag{8}$$

In the second case we reserve resources for each aggregate demand independently in link 3, even though all the sources will be eventually aggregated in the same traffic class in link 3, (this is referred sometimes as *isolation*), that is:

$$C_3 = eff(\{S_1, ... S_M\}, PLR_3) + eff(\{S_{M+1}, ... S_N\}, PLR_3) \tag{9}$$

In the following figures, the target packet loss rate for link 3 is set to 0.01 and 0.001. PLRb corresponds to the incurred packet loss rate from link 3 when using eq. 8 and PLRa corresponds to the incurred packet loss rate when using eq. 9.

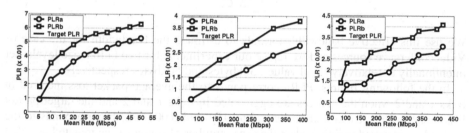

Fig. 4. Incurred PLR for *VoIP* (left), *Videoconference* (center) and *mixed* (right) traffic sources for link 3 with target PLR 0.01

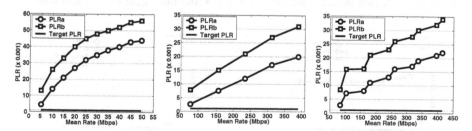

Fig. 5. Incurred PLR for *VoIP* (left), *Videoconference* (center) and *mixed* (right) traffic sources for link 3 with target PLR 0.001

As it can be seen, the effects of the traffic profile deformation, even one hop away from the node where the sources are firstly multiplexed, can lead to severe violations of the packet loss rate. Even if the *isolation* method (eq. 9) is used, which, given the form of the effective bandwidth formula of eq. 1, leads to more conservative re-sources reservation compared to when the two aggregate demands are merged into one bandwidth allocation (eq. 8), the target packet loss rate can be violated by more than one order of magnitude. Furthermore, the increase in the level of aggregation can lead to higher packet loss rate violations. This indicates that the detrimental effects of the aggregate traffic profile deformation can far exceed the positive effects due to the additional multiplexing gains that this increase is expected to have.

4.3 Scenario III: Bandwidth Allocation with Buffered Multiplexing Models

In order not to restrict ourselves to the bufferless approach, we repeat part of the above simulations using the buffered approach introduced in [7]. According to [7], for a source of type i with average rate m_i, the effective bandwidth is given by:

$$C_i = m_i + \delta \gamma_i /(2B) \tag{10}$$

where B is the buffer size, γ_i is the index of dispersion and $\delta = -\ln(\varepsilon)$, with ε the allowed loss rate. For M different types of traffic sources, with N_i sources of type i the total effective bandwidth is given by:

$$C = \sum_{i=1}^{M} N_i C_i \tag{11}$$

We use this effective bandwidth formula to estimate the bandwidth for the case of Videoconference sources and we repeat our experiments for two buffer size levels, set intentionally to relatively small values, 30kbytes and 50kbytes respectively.

In the following figures, PLRa corresponds to the average packet loss incurred in links 1 and 2 when the sources are aggregated in links 1 and 2 and the target packet loss rates of links 1 and 2 are set equal to 0.01 and 0.001. PLRb corresponds to the packet loss incurred in link 3 when the same number of sources is aggregated in link 3 with the capacities of links 1 and 2 set equal to the sum of peak rates of the sources they are carrying -links 1 and 2 are transparent with respect to packet loss- and the bandwidth allocated in link 3 set for target link 3 loss rate equal to 0.01 and 0.001.

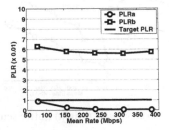 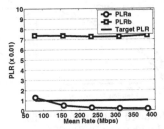

Fig. 6. Incurred PLR for *Videoconference* sources for link PLR 0.01 and queue size 30kbytes (left) and 50kbytes (right)

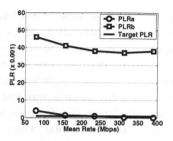 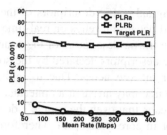

Fig. 7. Incurred PLR for *Videoconference* sources for link PLR 0.001 and queue size 30kbytes (left) and 50kbytes (right)

As it can be seen from PLRa, using eq. 11 for the sources on links 1 and 2 for target link PLR 0.01 gives the expected results for the incurred packet loss rate on these links, whereas it is quite optimistic for the case of target link PLR 0.001, leading to packet loss rate violations for a small numbers of multiplexed sources. Moreover, applying eq. 11 to calculate the effective bandwidth in link 3 (PLRb) can lead to excessive violation of the packet loss rate, especially for link 3 target loss rate 0.001. Furthermore, contrary to scenario II, the packet loss in link 3 does not increase with the level of aggregation. This is due to the additive nature of eq. 11, which becomes more conservative for increasing levels of aggregation and, therefore, can partly compensate for the detrimental effects of the aggregate traffic profile deformation.

5 Discussion

In this section we will first elaborate on the implications of our study and we will point out relevant issues that are raised. Subsequently, we will present some possible practical traffic engineering solutions for dealing with these issues.

5.1 General Implications

The first implication of our study is that the original traffic descriptors for a set of real-time traffic sources multiplexed at a network edge, are not only invalid at downstream nodes but, also, the incurred traffic profile deformation has, in general, a negative effect. That is, the traffic characteristics of a set of sources become, on average, worse in downstream nodes, which means that using the original traffic descriptors to depict the behavior of the sources can be an overly optimistic approximation.

The second implication is that for a set of sources, the greatest multiplexing gains are achieved at the network edge, where the sources are uncorrelated. This is clear from the increasing packet loss rates incurred in core links compared to those in edge links for the same bandwidth allocation scheme and number of multiplexed sources.

5.2 Implications for Admission Control

Regarding admission control, the results have certain implications on the effectiveness of an admission control scheme deployed in core nodes of a class-based IP network. If a TDAC scheme is deployed in a core node, it may fail if it is based on the original traffic descriptors of the traffic sources. An MBAC scheme is less likely to fail because it relies on real-time measurements of the aggregate traffic and uses only the traffic descriptors of the source requesting admission. However, the original traffic descriptors declared by the source requesting admission may not depict its behavior in core nodes. Due to multiplexing and buffering, the packets belonging to a source may arrive at an interface at a rate even exceeding the source's peak rate. That means that even the source's declared peak rate may not depict its worst-case behavior in core nodes [19]. Therefore, even a conservative MBAC scheme, which while deriving the admission control decision, assumes that the source requesting admission will be transmitting at its peak rate, may fail when applied in core nodes.

A similar problem for TDAC and MBAC schemes arises when performing admission control for inter-domain traffic; that is traffic that traverses transit domains. In this case, if an upstream domain submits the original traffic descriptors of the sources to a downstream domain (without taking into account the traffic profile deformation for the sources within this upstream domain) and the downstream domain performs admission control based on those traffic descriptors, this may lead to QoS violations.

In order for any admission control scheme that uses some kind of traffic descriptors to be reliable, when used in nodes other than the first multiplexing point, it should appropriately modify the traffic descriptors to depict the behavior of the sources at that specific multiplexing point. However, this is not trivial, since it requires the estimation of delay variation and induces per flow states [18].

Even if the effects of the traffic profile deformation can be taken into account and appropriate signaling methods exist to learn the sources behavior at downstream nodes, if a TDAC or MBAC scheme is applied on a link-by-link basis, if the link packet loss rates are not set so that the total end-to-end packet loss requirement of the flows traversing the larger number of links is satisfied, this will result in higher flow blocking probabilities for the flows traversing large number of links (long flows). This effect is similar to the discrimination against long flows in the case of EAC schemes [10]. However, if the link packet loss requirements are set so that the total packet loss requirement of long flows is satisfied, this can lead to underutilization of the resources for the links carrying short real-time traffic flows only, because in these links the per-link packet loss rates will be set to lower values than what is actually needed and the lower the target packet loss, the lower the achieved utilization [9].

5.3 Implications for Bandwidth Allocation

Regarding bandwidth allocation, the results suggest that if, during the network provisioning phase, the packet loss requirement is translated in a hop count constraint and the bandwidth allocation in core nodes is based on an effective bandwidth formula, even if done based on *edge-to-edge isolation*, that is that traffic aggregates multiplexed in the same traffic class in the core network are allocated resources on an ingress-egress pair basis, the consequences may be detrimental. Furthermore, similar to the admission control case, the capacity dimensioning should be done for link packet loss rates able to satisfy the end-to-end packet loss requirement of the longer flows, which will lead to underutilization of resources on links carrying only short flows.

5.4 Possible Practical Traffic Engineering Solutions

Part of the aforementioned issues regarding bandwidth allocation and admission control schemes (e.g., unfairness against long flows, underutilization of resources for links carrying only short flows) can be overcome if more sub-classes are configured and engineered in order to support the real-time traffic flows. This way, real-time traffic flows (with the same end-to-end packet loss rate requirement) can be aggregated in different classes so that in every link, only flows with similar target link packet loss rate requirements are aggregated in the same class. However, this would mean increasing the number of classes that must be engineered and supported in the routers. Apart from the added complexity in network dimensioning, increasing the number of classes that routers must support can lead to decreased forwarding performance [20].

A unified approach for bandwidth allocation and admission control that can be used to overcome the aforementioned problems, including the traffic profile deformation, is to apply admission control only at the network ingress and further downstream treat the real-time traffic aggregates in a peak rate manner. This simplifies network dimensioning, since it removes the packet loss related hop count constraint, and is feasible since the edge links are the most probable congestion points of a domain [21], whereas backbone links are overprovisioned [22]. This approach does not induce any states in the core network and does not require core routers to be aware of any signaling, which is desired for scalability and resilience reasons, and it is also proven to be very resource-efficient if resilience against network failures is required [23]. Furthermore, as our results suggest, the greatest multiplexing gains are achieved at the network edge, and by employing this approach, since losses are not incurred by the core network but are restricted to those incurred by the edge links, the target edge link packet loss rates can be set higher, which means increased utilization of these links.

6   Related Work

As aforementioned, the problem of traffic profile deformation has been verified in the past [8, 18, 19] and a number of solutions for dealing with it have been proposed.

One solution is the use of traffic-descriptor conserving scheduling disciplines in all links along the end-to-end paths. Example of such schedulers is the Rate-controlled Static Priority (RCSP) scheduler [24]. This preserves the original traffic descriptors of each flow going through it and provides zero packet loss guarantees. However, it requires per-flow queuing and keeping the traffic descriptors of each flow in each node. Furthermore, since RCSP is intended to provide zero packet loss guarantees, the deployment of RCSP can lead to unnecessarily conservative usage of network resources.

In [8] the issue of traffic profile deformation is discussed in the context of Constant Bit Rate sources in Asynchronous Transfer Mode networks and a solution for accounting for the traffic profile deformation of individual sources is given based on the estimation of delay variation, which, however, induces per-flow states and requires an appropriate method in order to obtain this delay variation estimation (e.g. in [25] the delay variation is estimated by employing a probing procedure). Similar to RCSP, this method may not be feasible in class-based networks since it can impose the requirement for adding functionality and keeping per-flow state in core nodes.

Contrary to these works, we focus not on the traffic profile deformation of individual sources, which induces per-flow states and can require added functionality in core nodes for the delay variation estimation. Instead, we focus on quantifying the aggregate traffic profile deformation for a variety of real-time traffic sources multiplexed in the same class. We discuss issues that are raised and how they can be addressed, without imposing the requirement for keeping per-flow state information in core nodes.

7 Concluding Remarks

In this paper we highlight several issues with respect to bandwidth allocation and admission control for the support of real-time traffic in class-based IP networks. We discuss the implications of topological placement, that is the location of the employed bandwidth allocation and admission control schemes with respect to the end-users and the various network boundaries (access, core, etc.), and we show that their performance depends on it; that is their performance at different points in a class-based IP network and for the same traffic load can be different and deviate greatly from the expected targets. Through simulations we also provide a quantitative view of these deviations. Finally, we propose a unified approach for bandwidth allocation and admission control that can overcome the detrimental effects of this "location-awareness".

References

1. W. Sun, P. Bhaniramka and R. Jain, "Quality of Service using Traffic Engineering over MPLS: An Analysis", IEEE LCN 2000.
2. G. Schollmeier and C. Winkler, "Providing Sustainable QoS in Next-Generation Networks", IEEE Communications Magazine, June 2004.
3. J. Lakkakorpi, O. Strandberg and J. Salonen, "Adaptive Connection Admission control for Differentiated Services Access Networks", IEEE Journal on Selected Areas in Communications, October 2005.
4. T. Bonald, A. Proutiere and J. Roberts, "Statistical Performance Guarantees for Streaming Flows using Expedited Forwarding", IEEE INFOCOM 2001.
5. M. Mowbray, G. Karlsson and T. Kohler, "Capacity Reservation for Multimedia Traffics", Distr. Syst. Eng., 1998.
6. R. Guerin, H. Ahmadi, and M. Naghshieh, "Equivalent Capacity and its Application to Bandwidth Allocation in High-Speed Networks", IEEE Journal on Selected Areas in Communications, September 1991.
7. C. Courcoubetis G. Fouskas and R. Weber, "On the Performance of an Effective Bandwidths Formula", International Teletraffic Congress 1994.
8. J. Roberts, U. Mocci and J. Virtamo, "Broadband Network Teletraffic", Final report of action COST 242, Springer 1996.
9. T. Bonald, S. Oueslati-Boulahia and J. Roberts, "IP traffic and QoS control: the need for a flow-aware architecture", World Telecommunications Congress, September 2002.
10. L. Breslau, E. Knightly, S. Shenker, I. Stoica and Z. Zhang, "Endpoint Admission Control: Architectural Issues and Performance", SIGCOMM 2000.
11. M. Grossglauser and D. Tse, "A Framework for Robust Measurement-Based Admission Control", IEEE/ACM Transactions on Networking, June 1999.

12. D. Eun and N. Shroff, "A Measurement-Analytic Approach for QoS Estimation in a Network Based on the Dominant Time Scale", IEEE/ACM Transactions on Networking, April 2003.

13. S. Lima, P. Carvalho and V. Freitas, "Distributed Admission Control for QoS and SLS Management", Journal of Network and Systems Management, September 2004.

14. K. Fall and K. Varadhan, "The ns manual" (www.isi.edu/nsnam/ns/ns_doc.pdf).

15. C. Chuah, L. Subramarian and R. Katz, "Furies: A Scalable Framework for Traffic Policing and Admission Control", U.C Berkeley Technical Report, May 2001.

16. B. Maglaris, D. Anastassiou, P. Sen, G. Karlsson and J. Robbins, "Performance Models of Statistical Multiplexing in Packet Video Communications", IEEE Transactions on Communications, July 1988.

17. I. Mas, V. Fodor and G. Karlsson, "The Performance of Endpoint Admission Control Based on Packet Loss", QoFIS 2003.

18. K. Shiomoto, N. Yamanaka and T. Takahashi, "Overview of Measurement-based Connection Admission Control Methods in ATM Networks", IEEE Communication Surveys 1999.

19. H. Perros and K. Elsayed, "Call Admission Control Schemes: A Review", IEEE Communications Magazine, November 1996.

20. M. Torneus, "Testbed for Measurement Based Traffic Control", Master's Thesis, KTH IMIT, Sweden, June 2000.

21. V. N. Padmanabhan, L. Qiu and H. J. Wang, "Server-based inference of Internet Link Lossiness", IEEE INFOCOM 2003.

22. G. Iannaccone, M. May, and C. Diot, "Aggregate Traffic Performance with Active Queue Management and Drop from Tail", Computer Communications Review, July 2001.

23. M. Menth, "Efficient Admission Control and Routing for Resilient Communication Networks", PhD Thesis, Univ. of Wurzburg, July 2004.

24. H. Zhang and D. Ferrari, "Rate-controlled Static-Priority queuing", IEEE INFOCOM 1993.

25. M. Dabrowski, A. Beben and W. Burakowski, "On Inter-domain Admission Control Supported by Measurements in Multi-domain IP QoS Networks", IEEE IPS 2004.

An Integrated Network Management Framework for Inter-domain Outbound Traffic Engineering*

Mina Amin, Kin-Hon Ho, Michael Howarth, and George Pavlou

Centre for Communication Systems Research, University of Surrey, UK
{M.Amin, K.Ho, M.Howarth, G.Pavlou}@surrey.ac.uk

Abstract. This paper proposes an integrated network management framework for inter-domain outbound traffic engineering. The framework consists of three functional blocks (monitoring, optimization and implementation) to make the outbound traffic engineering adaptive to network condition changes such as inter-domain traffic demand variation, inter-domain routing changes and link failures. The objective is to keep the inter-domain link utilization balanced under any of these changes while reducing service disruptions and reconfiguration overheads. Simulation results demonstrate that the proposed framework can achieve better load balancing with less service disruptions and re-configuration overheads in comparison to alternative approaches.

1 Introduction

Outbound Traffic Engineering (TE) [1,2,3,4] which has become increasingly important and has been well studied, is a set of techniques for controlling traffic exiting a domain by assigning the traffic to the best egress points (i.e. routers or links). The general problem formulation of outbound TE is: given the network topology, BGP routing information and inter-domain Traffic Matrix (TM), determine the best Egress Point (EP) for each traffic demand so as to optimize the overall network performance [2]. Since inter-domain links are the most common bottlenecks in the Internet [2], optimizing their resource utilization becomes a key objective of outbound TE.

In practice, network conditions change dynamically, which can make fixed outbound TE solutions obsolete and subsequently cause some inter-domain links to become congested over time. One such dynamic change is inter-domain traffic variation, which is typically caused by changes in user or application behavior, adaptations from the TCP congestion control or even routing changes from other domains [5]. In addition to these traffic variations, transient and non-transient inter-domain peering link failures might occur. According to [7] transient inter-domain link failures are common events and their duration is less than a few minutes. Upon failure of a peering link, there may be a large amount of traffic shifted to other available EPs, potentially leading to congestion on these new serving EPs if they are not carefully chosen. In

---

* This work was undertaken in the context of FP6 Information Society Technologies AGAVE (IST-027609) and EMANICS Network of Excellence on the Management of Next Generation Networks (IST-026854) projects, which are partially funded by the Commission of the European Union.

A. Helmy et al. (Eds.): MMNS 2006, LNCS 4267, pp. 208–222, 2006.

theory, although it is possible to perform outbound TE based on the other proposals in the literature [2,3,4] whenever any of those changes occur, it may require huge computational overheads and a large number of EP re-configurations given that previous proposals have not considered the reduction of reconfiguration changes and overheads. This can lead to excessive service disruptions and is not practical. As a consequence, lack of TE solutions that react to those dynamic changes rapidly enough will leave the network *unmanaged*. It is thus the focus of this paper to make outbound TE more adaptive to fast-changing IP networks by taking into consideration practical network operation and management constraints such as time-efficiency, reconfiguration overheads and service disruptions.

In this paper, we propose an Inter-domain Outbound Traffic Engineering (IOTE) framework that consists of two re-optimization components: (1) Primary Egress Point (PEP) re-optimizer that is designed to manage dynamic traffic variation and routing changes. This component handles primary outbound TE which determines EP selection under Normal State (NS, i.e. no inter-domain link failure); (2) Secondary Egress Point (SEP) re-optimizer that is designed to manage inter-domain link failures. This component handles secondary outbound TE which determines EP selection under Failure States (FS, i.e. transient and non-transient inter-domain link failure). A time-efficient heuristic algorithm is proposed for each optimization component. The overall objective of the IOTE FRAMEWORK is, *in spite of dynamic changes in network conditions, to balance the loads among inter-domain links under both NS and FSs, while reducing reconfiguration overheads and service disruptions.*

To the best of our knowledge, there is no such an integrated network management approach like the IOTE FRAMEWORK that addresses primary and secondary outbound TE simultaneously. The authors in [6] propose a multi-objective outbound inter-domain TE re-optimization that handles changes of the traffic pattern or routing failures with a minimal burden on BGP. However, they do not consider network performance under transient inter-domain link failures. On the other hand, the authors in [8] propose an intra-domain TE solution that is robust to transient intra-domain link failures and argue that relying on reactive robust solutions may not be appropriate or even feasible, since quickly computing and deploying a new robust solution can be challenging especially in today's large networks. In a similar fashion, changing EP configuration dynamically to avoid a transient failure may not be a practical solution since there is not sufficient time for network operators to configure their networks before recovering from the transient failure. Instead, in order to avoid human configuration and achieve fast recovery from inter-domain link failures, we pursue a proactive robust approach to manage the transient inter-domain link failure through the pre-computation of SEPs.

We compare the performance of IOTE FRAMEWORK with two alternative strategies. The first strategy does not consider any PEP or SEP re-optimization at all, while the second only considers PEP re-optimization. In our simulation model, we generate a series of random events to be handled by the different strategies, attempting to emulate realistic changes in network conditions. These events include traffic variation, routing changes and transient and non-transient inter-domain link failures. Simulation results demonstrate that the IOTE FRAMEWORK has the following key advantages over the other two alternatives: (a) in spite of network condition changes, maintains a better load balancing on inter-domain links under both NS and FSs; (b) limits the service disruptions and reconfiguration overheads.

This paper is organized as follows. In section 2, we present the proposed IOTE FRAMEWORK in detail. Section 3 presents the optimization problem handled by the PEP and SEP re-optimization components. We detail the proposed heuristic algorithms in Section 4. Section 5 presents two alternative strategies for performance comparison. Then, we present our evaluation methodology and simulation results in Sections 6 and 7 respectively. Finally we conclude the paper in Section 8.

2 Inter-domain Outbound Traffic Engineering Framework

The proposed IOTE FRAMEWORK is illustrated in Figure 1. The key idea of the framework is to continuously monitor[1] the network conditions and, if some optimization triggering policies are met, initiate the PEP and SEP re-optimization components based on the latest network conditions. The PEP and SEP solutions are then finally configured in the network if some implementation policies are met. The framework comprises three functional blocks which we explain in detail:

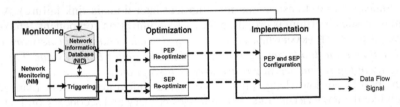

Fig. 1. Inter-domain Outbound Traffic Engineering Framework

1) **Monitoring Block:** It consists of Network Monitoring (NM) module, Network Information Database (NID) and triggering module. The NM module continuously monitors the network conditions to establish a global view of the network. The network information, which will be stored in the NID, includes inter-domain link utilization, overall traffic demand and BGP routing information. The authors in [9] presented a distributed management infrastructure that enables real-time views of network traffic to be generated. The key concept of their approach is that each router monitors its local resources (e.g. utilization of the attached links) and then stores the monitored data in local databases. When a real-time global view of network is needed for network management, the console system that is controlled by the network operator retrieves and processes the information from the databases at each router through specific query languages. To apply this distributed monitoring infrastructure to outbound TE, each egress router monitors the utilization of inter-domain links attached to it and collects the updated BGP routing information from the local Routing Information Base (RIB). On the other hand, each ingress router monitors the updated traffic

---

[1] In this paper, continuous monitoring can refer to 10 minutes interval according to [5,6]. However, there is a trade off between the accuracy of network conditions and monitoring overheads. In fact, the higher the accuracy of network conditions then the higher the monitoring overheads. The network operators may choose their best strategy to perform the network monitoring.

demand. Note that there are currently several hundred thousand prefixes in the Internet and collecting real-time changes for all the prefixes is challenging. As suggested in [1], TE can consider only a small number of prefixes that is responsible for large volume of traffic. As such, this monitoring block only needs to pay attention to these prefixes in order to significantly reduce the monitoring complexity as well as to make real-time data generation more efficient.

When the latest network conditions are known from monitoring, the NM module signals the triggering module. The triggering module invokes the re-optimization modules in the optimization block if some optimization triggering policies are met. The policy can be event-driven: re-optimization is invoked if an event occurs. In this paper, we use this event-driven policy for triggering the PEP and SEP re-optimizers as follows: (i) The PEP re-optimizer is invoked if the latest network utilization obtained by the monitoring exceeds a tolerance threshold α. This is a common policy since network providers often take actions to minimize congestion in their networks. Without loss of generality, in this paper we assume $\alpha = 50\%$ to be the borderline of congestion. In summary, the PEP re-optimizer aims to keep the network utilization under NS below α. (ii) The SEP re-optimizer is invoked when the network information database is updated by the NM. Note that, since the network may suffer from dramatically poor performance under FSs, keeping the SEP solution updated according to the changes is very important.

2) **Optimization Block:** it consists of PEP and SEP re-optimizers and requires as input the latest network information from the NID. The task of PEP re-optimizer is to re-assign the primary egress points to traffic flows under NS. The key objective is to achieve inter-domain load balancing while reducing reconfiguration overheads and service disruptions. The PEP re-optimizer is designed for managing dynamic traffic variation and routing changes. On the other hand, the task of SEP re-optimizer is to pre-compute a set of optimal secondary (i.e. backup) egress points for the traffic. Upon failure of an inter-domain link, the traffic affected by the failure will be shifted to the secondary egress points. The key objective is to achieve inter-domain load balancing under any single inter-domain link failure while reducing backup reconfigurations. The SEP re-optimizer is designed for managing inter-domain link failures. It is worth to mention that changing secondary egress points does not cause service disruption as the primary BGP routes remain intact. Details of the PEP and the SEP re-optimizers will be presented in Sections 3.1 and 3.2 respectively.

3) **Implementation Block:** it enforces the solutions produced by the PEP and SEP re-optimizer into the network according to the implementation policies. For the PEP re-optimization, the solution is enforced if it leads to better inter-domain load balancing than the previous configuration. On the other hand, a benefit-based implementation policy is used for the SEP re-optimization. The SEP solution is enforced if there is a significant gain in reducing the network utilization under FS compared to the previous attempt. The rationale of using this policy is to maximize the lifetime of the previous SEP solution in order to reduce frequent SEP reconfigurations. In fact, we avoid SEP reconfiguration until the latest solution provides a significant performance gain to the network. In this paper, we consider 10% performance gain as the significant gain for the SEP implementation policy. The current PEP and SEP configurations are updated in the NID in order to maintain the latest network information.

One way to implement the PEP and SEP solutions is to assign, for each prefix, the largest and the second largest value of BGP *local-pref* for the selected primary and secondary egress point respectively. To achieve faster failure recovery, the SEPs can be implemented by the proposal in [7] in which an IP tunnel is established to move traffic from the failed PEP to the pre-computed SEP.

3 Problem Formulation

Here, we present the optimization problem to be tackled by the PEP and SEP re-optimizer in the IOTE framework[2]. Table 1 shows the notation used in this paper.

Table 1. Notation used in this paper

NOTATION	DESCRIPTION
K	A set of destination prefixes, indexed by k
L	A set of egress points, indexed by l
S	A set of states $S=\{\varnothing U (\forall l \in L)\}$, indexed by s
I	A set of ingress points, indexed by i
$t(k,i)$	Bandwidth demand of traffic flows destined to destination prefix $k \in K$ at ingress point $i \in I$
$Out(k)$	A set of egress points that have reachability to destination prefix k
c^l_{inter}	Capacity of the egress point l
x^l_{sk}	A binary variable indicating whether prefix k is assigned to the egress point l in state s
u^l_s	Utilization on non-failed egress point l in state s. Its value is zero when $s=l$
$U_{max}(s)$	Maximum egress point utilization in state s
U_{worst}	Worst case maximum egress point utilization across all states
R,R'	Primary and secondary egress point reconfiguration limits
r_{PEP}, r_{SEP}	The number of actual primary and secondary egress point reconfigurations per re-optimization

3.1 Outbound TE PEP Re-optimization Problem Formulation

The PEP re-optimizer requires the following two items as input: (1) Network utilization: the latest utilization of each inter-domain link together with the maximum and the minimum; (2) The current possibly suboptimal PEP configuration: this includes the latest traffic matrix and BGP routing information. Note that the best EP for each destination prefix according to inter-domain BGP routing policy is known from the BGP routing information.

The task of the PEP re-optimizer is to re-assign the best EPs for destination prefixes, with the objective of balancing the utilization among inter-domain links under normal state ($s=\varnothing$) while reducing reconfiguration overheads and service disruptions. More specifically, the objective of inter-domain load balancing can be achieved by minimizing the maximum inter-domain link utilization. Moreover, inter-domain load

---

[2] In this paper, we focus the TE re-optimization objective on inter-domain resources due to the reason that capacity over-provisioning is usually employed by ISPs within their IP backbones [10]. In addition, since the objective of this paper is to demonstrate the principle of the outbound TE re-optimization, we consider only the single egress selection [2] as the outbound TE optimization problem. We leave the multiple egress selection as future work.

balancing and reducing EP changes (i.e. reconfigurations) are contradictory objectives: increasing the number of EP changes can improve inter-domain load balancing. In addition, balancing their trade-off is non-trivial. We therefore resort to using the ϵ-*constraint* method [11], which is one of the most favored methods of generating compromising bi-objective solutions. According to the ϵ-constraint method, the performance of an objective is optimized, while the other one is constrained so as not to exceed a tolerance value. Hence, we choose to place a constraint on the number of EPs changes that may be attained by the PEP re-optimization while minimizing the maximum inter-domain link utilization. Therefore, the optimization problem to be tackled by the PEP-re-optimizer can be formulated with the objective:

$$Minimize\ U_{max}(\varnothing) = Minimize\ Max_{\forall l \in L}(u_\varnothing^l) = Minimize\ Max_{\forall l \in L}(\frac{\sum_{k \in K}\sum_{i \in I} x_{\varnothing k}^l t(k,i)}{c_{inter}^l}) \tag{1}$$

subject to the following constraints:

$$r_{PEP} \leq R \tag{2}$$

$$\forall k \in K: \sum_{l \in Out(k)} x_{\varnothing k}^l = 1 \tag{3}$$

$$\forall l \in L, k \in K : x_{\varnothing k}^l \in \{0,1\} \tag{4}$$

Constraint (2) ensures that the number of EP changes does not exceed the limit R. A method used in this paper to determine R is presented in section 6.4. Constraints (3) and (4) ensure that only one EP is selected for each destination prefix as the PEP.

3.2 Outbound TE SEP Re-optimization Problem Formulation

The SEP re-optimizer requires as input the current SEP configuration as well as those inputs required by the PEP re-optimizer. The task of the SEP re-optimizer is to re-assign secondary egress points for destination prefixes, with the objectives of minimizing the worst case maximum inter-domain link utilization across all FSs (we assume single inter-domain link failure) while reducing secondary egress point changes. Similar to the PEP re-optimizer, we place a constraint on the number of secondary egress point changes while minimizing the worst case maximum inter-domain link utilization. Therefore, the optimization problem in the SEP re-optimizer can be formulated with the objective:

$$Minimize\ U_{worst} = Minimize\ Max_{\forall s \in S} U_{max}(s) \tag{5}$$

$$where\ \forall s \in S : U_{max}(s) = Max_{\forall l \neq s}(u_s^l) = Max_{\forall l \neq s}(\frac{\sum_{k \in K}\sum_{i \in I} x_{sk}^l t(k,i)}{c_{inter}^l}) \tag{6}$$

subject to the following constraints:

$$r_{SEP} \leq R' \tag{7}$$

$$\forall k \in K, s \in S: \sum_{l \in Out(k)} x_{sk}^l = 1 \tag{8}$$

$$\forall l \in L, k \in K, s \in S : x_{sk}^l \in \{0,1\} \tag{9}$$

$$\forall l \in L, k \in K \quad if \quad x_{\varnothing k}^l = 1 \quad then \begin{cases} x_{sk}^l = 1 & \forall s \in S/\{l\} \\ x_{sk}^l = 0 & \forall s = l \end{cases} \tag{10}$$

The term $x_{sk}^l t(k,i)$ consists of flows that are assigned to EP l as their PEP and also flows that are assigned to EP l as their SEP. Constraint (7) ensures that the number of SEP changes does not exceed the limit R'. A method used in this paper to determine R' is presented in section 6.4. Constraints (8) and (9) are equivalent to constraints (3) and (4), ensuring that only one EP is selected for each destination prefix as the SEP under each FS. Constraint (10) ensures that if prefix k is assigned to EP l under NS, then this prefix remains on l for all FSs except when the current FS is the failure on l. Note that, in comparison to the PEP re-optimization that minimizes the maximum link utilization only under NS, the SEP re-optimization optimizes the worst case maximum link utilization across all the FSs as expressed by the objective function (5).

4 Proposed Heuristics

PEP Re-optimization Heuristic: Local search algorithms have been shown to produce good results for many combinatorial optimization algorithms. We therefore propose an iterative local search algorithm for our heuristic as the following steps:

Step 1. Set r_{PEP} to zero and identify EPs with the maximum and minimum utilization $(U_{max}(\varnothing), U_{min}(\varnothing))$.

Step 2. Among all the prefixes whose PEP is the EP with maximum utilization $(U_{max}(\varnothing))$, search for the prefix that by reassigning it to the EP with minimum utilization $(U_{min}(\varnothing))$ would minimize the maximum EP utilization according to objective function (1). Re-assign the prefix to that EP, update both values of $U_{max}(\varnothing)$ and $U_{min}(\varnothing)$, and set $r_{PEP} = r_{PEP} + 1$.

Step 3. Repeat step 2 until either r_{PEP} reaches the limit R or there is no obvious performance improvement for $U_{max}(\varnothing)$ compared to the previous iteration. We define the threshold of obvious performance improvement to be 5%.

SEP Re-optimization Heuristic: Similar to the PEP re-optimization heuristic, we also propose an iterative local search algorithm for our SEP heuristic as follows:

Step 1. Set r_{SEP} to zero, calculate the maximum EP utilization under each potential FS

Step 2. Identify the EP l' with the worst case maximum link utilization U_{worst} under all FSs (i.e. the link with the highest $U_{max}(s)$ for all FSs). Calculate the utilization of EP $l^\wedge$ with the minimum link utilization $(U_{min}(s))$ for the state when l' has the maximum utilization.

Step 3. Among all the prefixes whose SEP is l', search for the prefix that by re-assigning it to $l^\wedge$ within that state would minimize the worst case maximum EP utilization according to objective function (5). Re-assign the prefix to $l^\wedge$, update both values of $U_{max}(s)$ and $U_{min}(s)$, and set $r_{SEP} = r_{SEP} + 1$.

Step 4. Repeat step 2 to 3 until either r_{SER} reaches the limit R' or there is no obvious performance improvement for the worst case performance compared to the previous iteration. We define the threshold of obvious performance improvement to be 5%.

5 Alternative Strategies

In this section, we present two alternative outbound TE strategies.

NO-REOPT: In this strategy neither the PEP nor the SEP re-optimization is considered. Therefore, in spite of any changes, the current configurations are always used.

PEP-REOPT-ONLY: This strategy only considers the PEP re-optimization. Therefore, in case of an EP failure (transient or non-transient), the affected traffic will be shifted according to the current SEP configuration. In comparison to the NO-REOPT, this strategy attempts to reactively improve the network performance under non-transient FSs, if the latest network performance obtained by the monitoring violates the threshold criterion. In fact, the PEP re-optimization is triggered to minimize the maximum EP utilization under the particular FS (i.e. in this special case that EP l has failed we have $s=l$ instead of $s=\varnothing$). Obviously, this strategy cannot improve the network performance in case of a transient failure due to the very short duration of the failure.

6 Evaluation Methodology

6.1 Network Topology and Inter-domain Traffic Matrix

Our experiment is performed on a topology with 30 egress routers, each being associated with an inter-domain link. We assume the capacity of all the inter-domain links to be identical. As suggested in [1], TE can focus only on a small fraction of destination prefixes that is responsible for a large fraction of traffic. Hence, we consider 4000 such prefixes in this paper. In fact, each of them may not merely represent an individual prefix but also a group of distinct destination prefixes that have the same set of candidate EPs [12] in order to improve network and TE algorithm scalability. Hence, the number of prefixes we consider could actually represent an even larger value of actual prefixes. We assume that, in the initial network condition, each EP acknowledges reachability to all the considered destination prefixes.

We generate a synthetic inter-domain traffic matrix for our evaluation. The traffic matrix consists of a set of inter-domain traffic flows that originates from each ingress point towards each of the considered destination prefixes. Each inter-domain traffic flow is associated with a randomly generated bandwidth demand according to uniform distribution. We remark that our traffic matrix generation is just our best attempt to model inter-domain traffic, as no synthetic model for actual behavior of traffic in real networks can be found in the literature.

6.2 Performance Metrics

The following metrics are used in our evaluation. For all these metrics, lower values are better than high values. The last two metrics are the re-optimization cost metrics.

- **The maximum EP utilization:** This refers to both $U_{max}(\varnothing)$ under NS and the $U_{max}(s)$ under FS s in objective functions (1) and (6) respectively.
- **Service Disruption per re-optimization:** A traffic flow (service) is disrupted if it is shifted to another EP due to re-optimization. We represent this

metric by SD and calculate it by adding the volume of all traffic flows disrupted for the PEP re-optimization.

- **The number of actual PEP and SEP reconfigurations per re-optimization:** These refer to r_{PEP} in (2) and r_{SEP} in (7) respectively.

6.3 Generated Events

Since no realistic model has been investigated for changes in network conditions, such as traffic variations, routing changes, inter-domain transient failures (TF) and non-transient failures (NTF), we generate a series of random events that attempts to emulate those realistic changes by assigning an occurrence probability to each event. By summarizing several relevant findings in [1,5,7,8], we consider TFs to be the most common event [7]. Hence, we assign the highest probability for TFs. The second highest probability is then assigned for traffic demand variations due to possibly frequent changes of user demands [5]. The lowest probabilities are assigned for routing changes and NTFs due to the stable nature of popular prefixes [1] and rare possibility of fiber-cut which is responsible for NTFs [8]. The performance of all the strategies under these events is investigated in section 7.

6.4 Determination of PEP and SEP Reconfiguration Limits

As mentioned earlier, minimizing the maximum EP utilization and reducing EP reconfigurations are contradictory objectives: the larger the number of EP reconfigurations, the better the expected value of the objective functions (1) and (5). To balance this tradeoff we calculate the PEP and the SEP re-configuration limits to restrict the negative effects brought from reconfiguration such as service disruptions and overheads while keeping the objective functions as low as possible.

Figures 2(a)-(b) illustrate the maximum EP utilization under NS and the worst case maximum EP utilization across all FSs respectively as a function of the number of actual EP reconfigurations. In Figure 2(a), the leftmost point on the curve represents a suboptimal PEP selection with maximum EP utilization under NS equal to 50%. In fact, since we chose the PEP re-optimization threshold to be 50% ($\alpha = 50\%$), we generated a suboptimal PEP selection solution with 50% maximum EP utilization under NS. Then this suboptimal PEP selection solution is improved by the IOTE FRAMEWORK using the PEP re-optimization heuristic without considering any reconfiguration constraint. The knee on the curve at point (41.2, 100) shows that only a very small load balancing improvement can be attained beyond 100 PEP reconfigurations. Hence, we set the PEP reconfiguration limit (R) to 100 and use the current PEP setting as an input for the SEP reconfiguration limit. In Figure 2(b), the leftmost point on the curve represents a suboptimal SEP selection with the worst maximum EP utilization across all FSs equal to 82.4%. Then this suboptimal SEP selection solution is improved by the IOTE FRAMEWORK using the SEP re-optimization heuristic without considering any reconfiguration constraint. The steepness of the left part of the curve indicates that large improvements in load balancing under FSs can be attained without large increase in the SEP reconfiguration. The knee on the curve at 400 SEP reconfigurations shows that only a very small improvement can be attained beyond this value. Hence, we set the SEP reconfiguration limit (R') to 400.

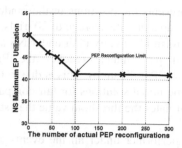

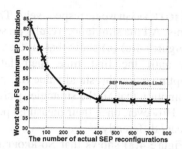

Fig. 2(a). PEP Reconfiguration limit

Fig. 2(b). SEP Reconfiguration limit

7 Simulation Results

7.1 Evaluation of the Maximum EP Utilization

We investigate the performance of all the strategies under several random events generated based on their occurrence probabilities. The randomly generated events are shown in Figure 3 and have the following order: (1) a sub-interval of small traffic fluctuations together with 5 TF. This sub-interval corresponds to positions $[0\ 3)^3$; (2) a sub-interval of gradual traffic increase together with 3 TF, 1 NTF and 1 TF. It corresponds to positions [3 10); (3) a sudden downward traffic surge, corresponds to position 10; (4) a sub-interval of small traffic fluctuations together with 5 TF. It corresponds to positions (10 13.7); (5) sudden routing changes, corresponds to position 13.7; (6) a sub-interval of small traffic fluctuations together with 7 TF. It corresponds to positions (13.7 18); (7) a sudden downward traffic surge, corresponds to position 18; (8) a sub-interval of gradual traffic increase with 1 TF, 1 NTF and 4 TF. It corresponds to positions (18 23.1); 9) sudden routing changes, corresponds to position 23.1; (10) a sub-interval of gradual traffic decrease with 6 TF. It corresponds to positions (23.1 27.1); (11) a sudden downward traffic surge corresponds to position 27.1 and finally (12) a sub-interval of small traffic fluctuations together with 6 TF. It corresponds to positions (27.1 30]. In addition, Figures 4(a), (b) and (c) show the maximum EP utilization under NS and FS *s* achieved by No-REOPT, PEP-REOPT-ONLY

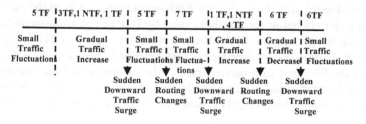

Fig. 3. Randomly generated events

---

[3] Note that, we will be using interval notations for the remainder of this section. In this notation, a "[" or "]" indicates that the number is inclusive, while a "(" or ")" indicates that the number is exclusive.

and IOTE FRAMEWORK respectively. The x axis represents the positions of the random events. All the simulation results presented in this paper are the average of 20 trials.

Figures 4(a)-(c) show that all the strategies perform identically both under NS and FSs until the first time their latest measured performance reaches the PEP re-optimization threshold value (i.e. 50% maximum EP utilization). This is due to our assumption that all the strategies start with the same initial solutions for fair comparison. However, once their measured performance violates the threshold value they start to react differently.

Figure 4(a) shows that NO-REOPT is the worst performer under all the events and cannot keep the maximum EP utilization under NS below the threshold value and its maximum EP utilization under FSs has dramatically poor performance. This phenomenon was expected due to the fact that neither PEP nor SEP re-optimization is considered in this strategy and the initial PEP and SEP solutions become less appropriate for the subsequent changes in the network conditions such as accumulation of traffic matrix variations and routing changes.

In contrast, Figure 4(b) shows that PEP-REOPT-ONLY can keep the maximum EP utilization under NS below the threshold value by the PEP re-optimization heuristic[4]. However, since no SEP re-optimization is considered in this strategy, its maximum EP utilization under FSs becomes poor and gets worse after subsequent events. Nevertheless, the overall FS network performance degradation in PEP-REOPT-ONLY is less severe compared to NO-REOPT. This result was expected since NO-REOPT does not apply any optimization heuristic as a result the failure of a congested EP and the assignment of its traffic flows over the non-optimized SEP may result in the assignment of a large number of traffic flows over already congested EPs which can cause a huge performance degradation. Whereas, in PEP-REOPT-ONLY as a result of an EP failure and the assignment of its flows over the non-optimized SEP does not lead to that much performance degradation due to the fact that the EPs are balanced under NS by PEP re-optimization. Moreover, PEP-REOPT-ONLY improves the maximum EP utilization when it exceeds the threshold value after NTFs. In total, Figure 4(b) shows 4 PEP re-optimizations to improve the maximum EP utilization after the traffic variations and routing changes and 2 PEP re-optimizations to improve the maximum EP utilization after the 2 NTFs which results to overall 6 PEP re-optimizations.

However, Figure 4(c) shows that the IOTE FRAMEWORK can keep the maximum EP utilization under NS below the threshold value by PEP re-optimization and moreover, can improve the maximum EP utilization both for TFs and NTFs by SEP re-optimization. In fact, its FS worst case performance is respectively 44% and 20% better than the FS worst case performance of the NO-REOPT and the PEP-REOPT-ONLY. Note that in the IOTE FRAMEWORK the maximum EP utilization in FSs is proactively re-optimized for both TFs and NTFs, whereas in the PEP-REOPT-ONLY, there is no

---

[4] Note that in PEP-REOPT-ONLY and IOTE FRAMEWORK, the maximum EP utilization under NS might exceed the tolerance threshold due to sudden changes. Nevertheless, both strategies are able to minimize the utilization below the tolerance threshold after the PEP re-optimization under the condition where there exist sufficient capacity to accommodate the latest overall traffic demands.

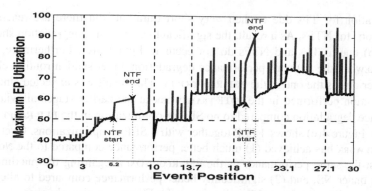

Fig. 4(a). Maximum EP Utilization of **No-Reopt** over event position

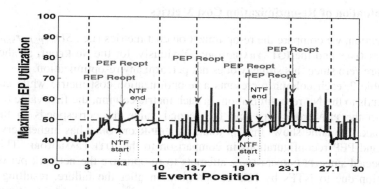

Fig. 4(b). Maximum EP Utilization of **PEP-Reopt-Only** over event position[5]

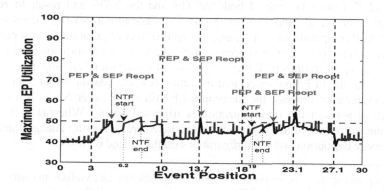

Fig. 4(c). Maximum EP Utilization of **IOTE-framework** over event position[5]

---

[5] Note that, in some cases, even though the maximum EP utilization violates the tolerance threshold, there is no re-optimization due to the reason that for those cases their re-configuration policy are not met.

re-optimization for TFs due to their very short duration[6] but there are reactive re-optimizations for NTFs. As a result, the significant performance degradation shown in Figure 4(b) due to TFs and NTFs do not occur in Figure 4(c). Furthermore, in the IOTE FRAMEWORK the network performance degradation under sudden routing changes is not as serious as the one in the PEP-REOPT-ONLY. This result was expected since SEP re-optimization performed in the IOTE FRAMEWORK at the earlier stages alleviates the performance degradation compared to no SEP re-optimization in the PEP-REOPT-ONLY.

In total, Figure 4(c) shows 4 PEP together with 4 SEP re-optimizations. Overall, the IOTE FRAMEWORK has achieved (1) much better performance compared to the NO-REOPT and almost the same performance as the PEP-REOPT-ONLY regarding the maximum EP utilization under NS, and (2) significantly better performance compared to alternative strategies regarding maximum EP utilization under FSs and routing changes.

7.2 Evaluation of Re-optimization Cost Metrics

In this section, we compare the re-optimization cost metrics (i.e.: SD, r_{PEP}, r_{SEP}) of the PEP-REOPT-ONLY and the IOTE FRAMEWORK. Obviously, for the NO-REOPT, all these cost metrics are zero since this strategy does not perform any re-optimization.

In Table 2, each column represents a re-optimization cost metric while each row corresponds to the Nth re-optimization. In each metric column, the first value (a) corresponds to the PEP-REOPT-ONLY and the second value (b) corresponds to the IOTE FRAMEWORK. The table shows that, in total, the PEP-REOPT-ONLY has higher service disruption and PEP reconfigurations in comparison to the IOTE FRAMEWORK. This result was expected since PEP-REOPT-ONLY attempts to re-optimize the network performance degradation due to NTFs by PEP re-optimization after the failure, resulting in two more PEP re-optimizations that corresponds to the 2nd and the 4th re-optimizations. Whereas in the IOTE FRAMEWORK the proactive SEP re-optimizations that correspond to the 1st and 3rd rows take care of both the TFs and the NTFs and result to zero re-optimizations on the 2nd and 4th rows. Moreover, since PEP-REOPT-ONLY does not perform any SEP re-optimization, it requires more PEP reconfiguration for re-optimizing the network performance after sudden routing changes which corresponds to the 3rd and the 6th re-optimizations in Table 2. Note that at these two re-optimizations the r_{PEP} have exceeded the PEP limit calculated in section 6.4. The reason is that after the first PEP re-optimization process, the maximum EP utilization under NS is still over the threshold, as a result the re-optimization is triggered again[7]. Whereas in the IOTE FRAMEWORK the proactive SEP re-optimizations take care of routing changes and result to less service disruption and re-configurations on the 3rd and 6th rows.

---

[6] If TF happens at the time of network conditions monitoring and violates the network performance threshold criterion, the PEP re-optimization is triggered. However, since the TF has a very short duration, it is recovered earlier than the configuration takes place. At this point network operator could simply ignore such re-optimization. In this paper, we assume that the network operator takes care of this task and therefore no re-optimization due to TFs are shown in the graphs.

[7] In fact, in Table 2 on the 3rd row the first value of r_{PEP} is the sum of two numbers of actual PEP re-configurations that correspond to the two consecutive PEP re-optimizations ($r_{PEP=}100+70=170$). Similarly, for the first value of r_{PEP} on the 6th row we have $r_{PEP=}100+40=140$.

In summary, the IOTE FRAMEWORK incurs almost 40% less service disruptions/PEP reconfigurations in comparison to the PEP-REOPT-ONLY at the cost of 4 SEP re-optimizations which result in 710 SEP reconfigurations, to keep the network perform-ance under FSs well balanced. We recall that the SEP reconfiguration does not cause service disruption. In addition, less service disruptions/PEP reconfigurations in our framework may imply better network stability compared to the PEP-REOPT-ONLY. Also, in our framework the numbers of actual PEP and SEP reconfigurations per re-optimization have never exceeded their limits.

Table 2. Re-optimization cost metrics for (a) **PEP-Reopt-Only** and (b) **IOTE framework**

RE-OPTIMIZATION	SD		r_{PEP}		r_{SEP}	
	a	b	a	b	a	b
1	9533	9533	70	70	0	150
2	5958	0	40	0	0	0
3	22641	11916	170	90	0	200
4	7150	0	40	0	0	0
5	9533	9533	60	60	0	160
6	19066	13108	140	100	0	200
Total	73881	44090	520	320	0	710

8 Conclusion

In this paper, we have addressed the problem of existing outbound TE solutions in case of dynamic changes in network conditions such as traffic variations, routing changes and inter-domain link failures. Hence, we have proposed an Inter-domain Outbound Traffic Engineering (IOTE) framework that aims to balance the load on inter-domain links under both normal and failure states, while reducing service dis-ruptions and reconfiguration overheads. We developed time-efficient heuristics to achieve the framework objectives and compared its performance to two alternative strategies. Our simulation results show that our proposed framework performs better compared to the alternative strategies regarding our objectives. We believe that our work provides insights to network operators on how to keep a balanced network espe-cially under transient and non-transient inter-domain failures in spite of traffic varia-tions and inevitable routing changes by limiting egress point changes.

Reference

[1] N. Feamster et al., "Guidelines for Interdomain Traffic Engineering," *ACM CCR*, October 2003.
[2] B. Bressound et al., "Optimal Configuration for BGP Route Selection," *IEEE INFOCOM*, 2003.
[3] S. Uhlig et al., "Interdomain Traffic Engineering with Minimal BGP Configurations," *ITC*, 2003.
[4] K. Ho et al., "Multi-objective Egress Router Selection Policies for Interdomain Traffic with Bandwidth Guarantees," Proc. *IFIP Networking*, 2004.

[5] R.Teixeira et al, "Traffic Matrix Reloaded: Impact of Routing Changes," Proc. *PAM*, 2005.

[6] S. Uhlig et al, "Designing BGP-based outbound traffic engineering techniques for stub ASes" *ACM SIGCOMM CCR,* October 2004.

[7] O. Bonaventure et al., "Achieving Sub-50 Milliseconds Recovery Upon BGP Peering Link Failures," Proc. *ACM CONEXT*, 2005.

[8] A. Sridharan et al., "Making IGP Routing Robust to Link Failures," Proc. *IFIP Networking*, 2005.

[9] K.S. Lim et al, "Real-time Views of Network Traffic using Decentralized Management," Proc. *IFIP/IEEE Integrated Network Management (IM),* 2005

[10] T. Telkamp, "Traffic Characteristics and Network Planning," *NANOG 2002.*

[11] V. Chankong et al, Multiobjective Decision Making-Theory and Methodology, Elsevier, 1983

[12] A. Broido et al., "Their Share: Diversity and Disparity in IP Traffic," Proc. *PAM*, 2004.

RTS/CTS Based Endpoint Admission Control for VoIP Over 802.11e

John Fitzpatrick, Seán Murphy, and John Murphy

Performance Engineering Lab,
UCD School of Computer Science and Informatics,
University College Dublin, Belfield, Dublin 4
John.fitzpatrick@ucd.ie

Abstract. In this paper an endpoint-based admission control mechanism for VoIP over WLAN is proposed. The mechanism operates by first measuring the utilisation of the channel and comparing this to a pre-determined threshold, if the channel utilisation exceeds this threshold, the call is rejected. One important aspect of the mechanism is that it can operate in the presence of hidden terminals. This is done by using the Request-to-Send/Clear-to-Send WLAN signalling to determine channel utilisation. The scheme is designed to be flexible; it can operate with heterogeneous data rates, with varying traffic types and in the presence of legacy 802.11 nodes. The scheme was developed and evaluated using the Qualnet network simulator. An empirical approach was used to determine appropriate admission thresholds. Then, simulations were performed to demonstrate the successful operation of the scheme.

Keywords: Call Admission Control, VoIP, 802.11e, RTS/CTS.

1 Introduction

WLAN and VoIP are two technologies that have seen unprecedented growth in recent times. The use of VoIP via WLAN access is still quite uncommon, however. The release of multimode terminals supporting applications such as Skype or Googletalk heralds a significant change for use of VoIP over wireless access, but there are still some issues to be addressed in the case of WLAN.

Unlike cellular systems, WiFi was not designed to transport delay-sensitive data. The legacy IEEE 802.11 MAC [1] uses a contention based mechanism that provides best effort throughput for all traffic. This mechanism is inadequate for supporting delay-sensitive, realtime applications such as VoIP; a one-way delay of no more than 150ms is required to deliver PSTN quality, although delays of up to 400ms can be tolerated [2]. Loss thresholds are somewhat codec dependent, but, in general, losses in excess of 3-4% are problematic.

To address this problem, the IEEE developed the 802.11e standard [3] to introduce QoS support to 802.11. A new contention mechanism called Enhanced Distributed Channel Access (EDCA) was developed as part of the standard. It provides QoS support by prioritising channel access for real time applications.

While EDCA can protect real time applications in the presence of best effort traffic, the presence of too many users generating high priority traffic can result in

A. Helmy et al. (Eds.): MMNS 2006, LNCS 4267, pp. 223–234, 2006.

unacceptable QoS for priority applications. For this reason, call admission control mechanisms are needed to prevent existing VoIP calls from being degraded due to the addition of further calls. If one more VoIP call than can be supported by the network is added, the quality of all existing calls becomes unacceptable.

The call admission control scheme proposed in this paper is an endpoint centric approach which requires no modifications to the WLAN infrastructure. More specifically, it does not require modifications to APs or routers in the network. The mechanism operates exclusively in the terminal and hence only an upgrade to software in the mobile terminal is needed to realise the scheme.

The admission control mechanism operates by measuring the utilisation of the radio channel. This is then compared with a pre-determined threshold. If the channel utilisation is below the threshold, the call is admitted; else the call is rejected.

To make accurate AC decisions the collective channel utilisation of all nodes must be correctly estimated. This can be problematic in the presence of hidden nodes[1]. To overcome this problem, it is assumed here that the RTS/CTS mechanism is used for transmission of all packets. While this does incur an extra overhead, the advantage of using this approach is that each node can detect activity generated by hidden nodes.

This paper is structured as follows. Section 2 provides a brief overview of some related work in the area of admission control for WLAN. Section 3 discusses WLAN technologies relevant to the proposed scheme. In section 4 the design and implementation of the proposed scheme is discussed. Section 5 describes the simulation setup and is followed by a discussion of the results obtained. This paper is then concluded in section 6.

2 Related Work

A large amount of the related work in the area of call admission control (CAC) over WLAN is access point (AP) centric. Most approaches require network infrastructure modifications such as the introduction of non-standard APs. Standardised solutions based on 802.11e have been considered but these fail to provide admission control for legacy 802.11a/b/g equipment.

2.1 Distributed Admission Control (DAC)

DAC, originally included in earlier drafts of 802.11e but later removed, was proposed by the IEEE 802.11e working group to protect active QoS streams such as VoIP and video in the EDCA mode of operation. The AP announces a transmission budget as part the beacon frame; the transmission budget specifies the amount of extra transmission time available for each access category (AC) over the next beacon interval. To calculate the transmission budget the AP measures the amount of utilised transmission time for each AC per beacon interval. The utilised transmission time is then subtracted from the transmission limit to obtain the transmission budget. This is done individually for each AC. From the transmission budget announced by the AP,

---

[1] The hidden node problem refers to a situation in which two or more nodes are in range of the access point (AP), but out of radio range of one another. Consequently the transmissions of multiple nodes may interfere due to simultaneous transmission, leading to a collision.

each station sets its own transmission budget for each AC based on the successfully used transmission time during the previous beacon interval. When the budget for an AC is depleted, no new streams will be allowed and all existing flows are prevented from increasing their transmission time. However, as previously stated, it fails to provide admission control for legacy equipment.

2.2 VMAC and VS

The Virtual MAC (VMAC) and virtual source (VS) proposed in [5, 6] are a set of algorithms that emulate the MAC and source applications in order to estimate achievable QoS. VMAC and VS operate in parallel with the real MAC and monitor the load on the radio channel to estimate service level quality metrics such as delay and loss. The VS consists of a virtual application and a virtual queue. The virtual application produces data packets just as a real application would, the packets are then placed in the interface queue. The VMAC performs scheduling, CSMA/CA and random back off just as a real MAC would. However no packet is transmitted, the VMAC estimates the probability of a collision if the packet was sent. A collision is detected when another node chooses the same time slot to transmit. In this case the VMAC enters a back off procedure as a real MAC would. The VMAC algorithm measures delay and loss on the uplink of the client. However, the limiting factor for a VoIP call is usually on the downlink as the AP which has to contend for the medium many more times than each client and therefore becomes a bottleneck. Another issue with this approach is that VMAC can only monitor the radio channel of nodes that are within radio range of each other and hence cannot give a true estimation of achievable QoS that would be delivered by the network due to the presence of hidden terminals.

3 Channel Access Mechanisms

This section gives an overview of DCF and EDCA channel access mechanisms as they are relevant to the proposed CAC scheme. Also, the RTS/CTS mechanism that allows the proposed scheme to operate in the presence of hidden terminals is also discussed.

3.1 Distributed Coordination Function (DCF)

The IEEE 802.11 MAC uses a contention based channel access mechanism called DCF. Although DCF is much written about and is well understood, this section is included for completeness.

DCF employs carrier sense multiple access with collision avoidance (CSMA/CA) to enable distributed channel sharing. Each DCF station with a packet to transmit must sense the wireless channel and determine it to be idle for a period equal to one distributed interframe space (DIFS) before proceeding with transmission. If the channel is determined to be busy, the station will wait until the channel becomes idle for a period defined by the DIFS. After detecting the channel to be idle for DIFS, the station waits a random backoff interval before attempting to transmit a frame.

DCF implements a binary exponential backoff to select the backoff interval. The backoff interval is uniformly selected over the interval [0, CW-1], where CW is the current contention window size in units of timeslots. CW has an initial value of CW_{min} for the first transmission attempt. After each unsuccessful transmission attempt the CW is doubled to a maximum value of CW_{max}. On receiving acknowledgement of a successfully received frame the CW is reset to CW_{min}. The backoff counter is decremented once every timeslot after the medium is idle for a DIFS period. If the medium becomes busy during a backoff interval the backoff counter is paused and resumes again after the medium becomes idle for a DIFS. A transmission is attempted when the backoff counter reaches zero.

In DCF each successfully received frame must be acknowledged by the receiving station. Upon successful reception of a frame the receiving station transmits an ACK frame after a period equal to the short interframe space (SIFS). The transmitting station assumes the frame to be lost if an ACK frame is not received within a specified ACK timeout, it then reschedules the frame to be retransmitted. As SIFS is smaller than DIFS, ACK frames will access the medium earlier than data frames, meaning they have priority. This means that other stations are prevented from accessing the medium during this short period before the ACK can be sent, as they must wait for at least DIFS before attempting any transmissions.

3.2 RTS/CTS

The CSMA/CA mechanism implemented by DCF suffers from the hidden node problem. To combat the hidden node problem, the request-to-send/clear-to-send (RTS/CTS) mechanism was developed. The RTS/CTS is a four way handshake mechanism designed to reduce the number of frame collisions. Before a data frame can be transmitted a successful handshake must take place between the sending and receiving nodes using RTS/CTS control frames. In general RTS/CTS is only used for large packets due to the signalling overhead the mechanism introduces. For small packets it is more efficient to perform a retransmission if a collision occurs.

Using RTS/CTS, the sending node transmits an RTS control frame containing its source address as well as the intended destination address and required channel time for the transmission of the data frame. Having successfully received the RTS frame, the destination node transmits a CTS frame after a SIFS. The CTS frame contains only the sending node address and the required channel time. This serves two purposes, it informs the sender to proceed with transmission of the data frame and it also notifies other nodes of the impending transmission duration so they will not attempt a transmission, essentially reserving the channel for the specified period to allow for the transmission of the data frame and corresponding ACK.

3.3 Enhanced Distributed Channel Access (EDCA)

The legacy 802.11 MAC provides distributed channel access shared equally among all users and traffic types, which works well for best effort traffic. However, the legacy DCF mechanisms cannot provide the QoS needed by real time delay sensitive applications such as VoIP. EDCA is a contention based channel access mechanism described in 802.11e, it provides an enhancement to DCF allowing for QoS support based on prioritized access to the medium. Prioritization is achieved through eight prioritizations (0-7) that further map to four access categories (AC) (0-3), with each access category obtaining differentiated channel access. Each packet is passed to the

MAC layer with a specific priority value specified in the IP packet header. The EDCA mechanism then maps the frame to a particular AC based upon the priority value. The mappings are shown in table 1.

Table 1. Priority & Access Category Mappings

Traffic Type	Priority	Access Category (AC)
Best Effort	0	0
Best Effort	1	0
Best Effort	2	0
Video	3	1
Video	4	2
Video	5	2
Voice	6	3
Voice	7	3

An AC within a station independently contends for the medium using a backoff process similar to DCF as previously discussed. Each AC has different parameters associated with the backoff mechanism, which give the different channel access prioritizations.

The AC dependent EDCA parameters are $AIFS[AC]$, $CW_{min}[AC]$, and $CW_{max}[AC]$, which replace the DCF parameters $DIFS$, CW_{min}, and CW_{max} respectively. $AIFS[AC]$ (Arbitration InterFrame Space) is used instead of $DIFS$ as the minimum time the channel must be sensed idle before a transmission can proceed.

As each AC maintains independent backoff counters, an issue known as a *virtual collision* can arise, in which two or more ACs within a single station finish backoff during the same timeslot. When this occurs the AC with the highest priority is allowed to transmit, and the other ACs double their CW values and recontend for channel access.

4 RTS/CTS Based Admission Control

In this section the proposed AC scheme is described. The essence of the proposed scheme is to use transmission duration information contained in the header of RTS/CTS control frames to estimate the current utilisation of the wireless channel. The channel utilisation estimate can then be used to make admission control decisions for VoIP calls. The use of RTS/CTS mitigates any hidden node problems and allows each station to passively monitor network activity.

4.1 Call Admission Decision

The call admission decision is performed by the application, utilising lower layer metrics. The decision to admit or reject the call is based upon whether the channel can support the additional call without adversely affecting calls already in progress.

Each call added to the network increases the channel utilisation by an amount dependent on the station's transmission rate. When the capacity of the channel is exceeded in the downlink direction, the end-to-end delay experienced by the VoIP calls rises dramatically, as shown in [8]. The increase in end-to-end delay greatly reduces the VoIP quality experienced by all users present in the network.

By using 802.11e real time traffic takes prioritisation over best effort. As was shown in [7], as the volume of high priority real time traffic increases; the throughput of low priority best effort traffic approaches zero. This allows the proposed scheme to operate in the presence of best effort traffic.

The admission decision uses predetermined channel utilisation threshold values, the determination of which will be discussed in the results section. If the channel utilisation as measured by the node when the call attempt is being made is greater than the threshold value, the call is rejected, else it is admitted. The threshold values chosen are the maximum channel utilisation values at which one extra VoIP call can be added to the network without excessively increasing the delay so as to reduce the QoS of the admitted and existing call. Each threshold value is chosen so as to allow one extra call to be added without overloading the network, such that when a node performs an admission control decision using the threshold values and the call is admitted it will be supported.

4.2   RTS/CTS Control Frames

Having successfully contended for the medium a station must transmit an RTS control frame to the intended recipient of the impending data frame. The transmission duration field of the control frame is calculated based upon the current physical layer transmission data rate, the data frame size and the basic data rate at which all control frames will be transmitted. Upon successful reception of an RTS frame the recipient responds with a CTS frame. The CTS header duration field contains the transmission duration specified in the RTS frame, less the time taken to transmit the RTS frame a SIFS period. In other words each header duration field of each control frame specifies how much longer the channel will be occupied for. This allows other stations to backoff until after the transmission duration has finished.

The proposed mechanism is intended for use in an infrastructure mode WLAN, hence all traffic is transmitted via the AP. Each station in the network monitors all RTS/CTS control frames originating from the AP. RTS frames originating from the AP will have a source address matching that of the AP. As CTS frames contain only a destination address, any CTS frames received whose destination address does not match that of the AP are assumed to be destined for other nodes in the same BSS and as such must have been transmitted by the AP (where infrastructure mode of operation is assumed).

Each stations MAC layer maintains a channel utilisation timer that is incremented using the transmission duration from the control frames sent by the AP. When an RTS frame is received from the AP, the channel utilisation timer is simply updated with the value specified in the duration field of the RTS frame. However, when a CTS frame is received from the AP the incremented time is calculated as the duration specified in the CTS header with the addition of the transmission duration of the RTS packet and a SIFS. This corrects for the smaller transmission duration specified by CTS frames than by RTS frames.

Every user definable period T, the channel utilisation is calculated as a percentage of total channel time that equals the period T over which the measurement is being performed. The utilisation timer is then reset for the next measurement period. AP beacon frames and other management traffic that do not use RTS/CTS control frames are ignored as the amount of channel time occupied by these is neglible.

4.3 Cross Layer Communication

In the proposed scheme, cross layer communication takes place between the application layer and both the MAC and physical layers. The admission decision procedure is performed at the application attempting to make a VoIP call. The MAC layer channel utilisation metric and the physical layer transmission data rate are passed to the application upon receipt of application layer triggers. These parameters are then be used to make the admission control decision. The data rate is needed to determine channel utilisation of the flow to be admitted, this will be discussed further in the next section.

5 Simulation Results

Implementation and simulations of the proposed CAC scheme were carried out using the Qualnet network simulator [9]. Development of the scheme required modification to multiple layers within Qualnet, in particular the 802.11 MAC and CBR application.

This section provides a brief overview of the simulation setup and a discussion of the results obtained.

5.1 Simulation Setup

The wireless environment was simulated using the 802.11e MAC layer and the 802.11g physical layer giving a maximum data rate of 54Mbps. To simulate VoIP traffic, constant bit rate (CBR) sources are used. Each CBR source provides a stream of packets that simulates simplex voice data. Full duplex traffic was modelled using two simplex traffic sources, one on the uplink and one on the downlink. The motivation behind simulating voice calls as full duplex was based on the most commonly used VoIP program Skype, which uses full duplex (i.e. no silence suppression is used)[2].

The VoIP codec used for the simulations was G.711 with a frame size of 20ms and a packet size of 160 bytes. This was chosen as it gives the highest voice quality of all standardised VoIP codecs. Lower bit rate codecs such as G.729 were also considered. However, it was found that using lower bit rate codecs did not offer any increase in capacity or throughput and simply reduced the maximum attainable quality of each call. This is due to the overhead introduced by the four way handshake of RTS/CTS, as each RTS/CTS control frame is transmitted at a low data rate.

---

[2] Skype uses full duplex for two reasons. Transmitting silent packets maintains UDP bindings at NAT (Network Address Translation). Also, if data is being transmitted over TCP the silent period packets prevent a reduction in the congestion window size during the silent period.

5.2 Determining the Channel Utilisation Thresholds

Simulations were performed to calculate the channel utilisation threshold values that are used by the application layer to make CAC decisions. The amount of channel utilisation a station absorbs is dependent on the physical data rate of the station. Stations in low data rate zones of the WLAN require more channel time to transmit their data and hence place more load on the network.

To calculate the threshold values individual experiments for each of the 802.11g supported data rates were performed. In each experiment a single VoIP call at the data rate being investigated was initially added to the system with subsequent 54Mbps calls being added incrementally until the capacity was exceeded. Each experiment was run for 1000 seconds with the channel utilisation being measured every 100ms. The mean and standard deviation of the channel utilisation was then calculated at the end of each simulation. Each data point in the presented results is the average of five such simulation runs.

The threshold value for each case could then be obtained by examining the channel utilisation as measured by each station as the additional calls were added. The threshold value for each data rate was chosen as the mean value plus twice the standard deviation of the channel utilisation two calls before capacity was exceeded. The mean channel utilisation value two calls before the capacity is exceeded, was used so that when a station measures the channel utilisation and it is found to be below the threshold value the additional call admitted will not exceed the system capacity.

Figure 1 and figure 2 shows the channel utilisation and average end-to-end delay for a varying number of 54Mbps calls. As can be seen the system capacity is 24 VoIP calls at the highest possible data rate of 54Mbps. When the 25th call is added to the system the end-to-delay increases dramatically and can not be supported[3].

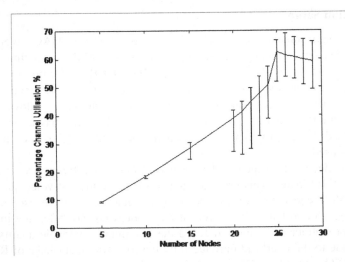

Fig. 1. Channel Utilisation for varying number of 54Mbps calls

---

[3] Note that 802.11g has be shown to accommodate a significantly higher number of VoIP calls, however this is in the case where RTS/CTS signalling is not used.

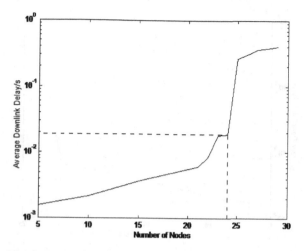

Fig. 2. End-to-End delay for varying number of 54Mbps calls

The mean channel utilisation for 54Mbps calls was found to be 48 percent, with a standard deviation of 3.5 percent. Hence, the threshold value of the mean plus twice the standard deviation was found to be 55 percent.

The same experiments were carried out for the possible 802.11g data rates, 54, 48, 36, 24, 12 and 6 Mbps. Although different data rates were used, it was found that not all data rates had different threshold values; only two differing threshold values were found to be needed. A threshold value of 55 percent channel utilisation was calculated for calls taking place in data rates greater than 12Mbps. The following figures 3 and 4 shows the channel utilisation and the delay for the 12Mbps case. Each data point in the figure 3 represents the average channel utilisation when there are X amount of calls present on the network with all calls at a data rate of 54Mbps with the exception of the final call in each case which is at a rate of 12Mbps.

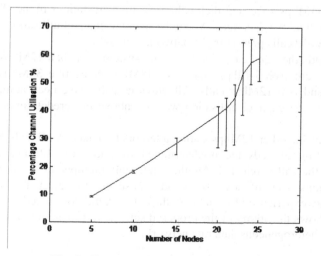

Fig. 3. Channel Utilisation (In the 12Mbps case)

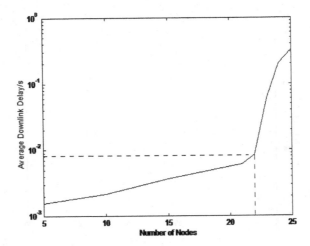

Fig. 4. End-to-End delay (In the 12Mbps case)

The same experiments were performed for the 6Mbps case and from the results obtained, the channel utilisation threshold value for calls with data rates at and below 12Mbps was found to be 49 percent, calculated using the mean value plus twice the standard deviation, as done previously.

5.3 Admission Control with Heterogeneous Data Rates

Figure 5 shows the proposed call admission control mechanism in operation. A simulation network is setup such that there are five stations in each data rate zone of the WLAN. Beginning at 500 seconds and in increments of 10 seconds, each station attempts to make a G.711 VoIP call. The simulation was configured so that all stations in the 54Mbps zone start their calls first, then all stations in the 48Mbps zone and so on through all the data rates. As can be seen the channel utilisation increases as the number of VoIP calls present in the network increases.

At 710 seconds the 22nd call is added to the system. This is a 12Mbps call and at this point there are five 54Mbps calls, five 48Mbps calls, five 36Mbps calls, five 24Mbps calls and two 12Mbps calls. All previous calls have been admitted by the CAC scheme as the channel utilisation was not above the predetermined threshold values.

At 720 seconds another 12Mbps station attempts to make a VoIP call, however the channel utilisation exceeds the predetermined threshold for 12Mbps calls of 49 percent, hence the call is rejected. Another eight call attempts are made by the eight remaining stations over the next 80 seconds, these too are rejected. The maximum end-to-end delay experienced by a node in the network did not exceed 0.0104 seconds (10ms). This shows the ability of the proposed scheme to provide QoS guarantees for VoIP calls in a heterogeneous data rate network.

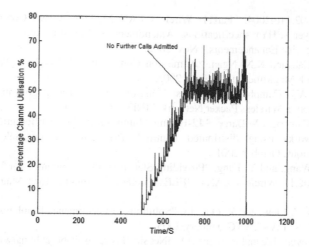

Fig. 5. Heterogeneous Data Rate Simulation

6 Conclusion

In this paper an endpoint based call admission control mechanism for VoIP over heterogeneous transmission rate IEEE 802.11g/e networks has been proposed. The scheme can operate in the presence of varying traffic types and with legacy 802.11 stations.

The CAC decision was based on an estimate of the current channel utilisation. This was determined using information transmitted in the RTS/CTS signalling messages. This was then compared to some predetermined thresholds. If the channel utilisation exceeded the threshold, then the call was rejected.

An empirical approach to determining the thresholds was used. These threshold values were then validated in a scenario with nodes connecting using different date rates.

Future work will be to further evaluate the operation of the scheme in the presence of different traffic types. Also the impact on the CAC decision of node mobility and varying calls arrival patterns needs to be assessed.

Acknowledgements

The support of the Irish Research Council for Science, Engineering and Technology (IRCSET) is gratefully acknowledged.

References

1. IEEE Std. 802.11, "Wireless LAN Medium Access Control (MAC) and Physical Layer (PHY) Specifications," 1999.
2. ITU-T, "Series G: Transmission Systems and Media Digital Systems and Networks" G.113, February 2001.

3. IEEE Std. 802.11e-2004, "Part11: Wireless LAN Medium Access Control (MAC) and Physical Layer (PHY) specifications. Amendment 8: Medium Access Control (MAC) Quality of Service Enhancements," Nov. 2005.
4. D. Gao, J. Cai, and K.N. Ngan, "Admission Control in IEEE 802.11e Wireless LANs," IEEE Network Magazine, July/August 2005.
5. M Barry, AT Campbell, A Veres "Distributed Control Algorithms for Service Differentiation in Wireless Packet Network," INFOCOM, 2001.
6. A Veres, A Campbell, M Barry, S Li-Hsiang, "Supporting service differentiation in wireless packet networks using distributed control," IEEE Journal on Selected Areas in Communications, October 2001.
7. H. Zhai, J. Wang, and Y. Fang, "Providing Statistical QoS Guarantee for Voice over IP in the IEEE 802.11 Wireless LANs" IEEE Wireless Communications Magazine, February 2006.
8. W. Wang, S. C. Liew, and V. O.K. Li, "Solutions to Performance Problems in VoIP over a 802.11 Wireless LAN", VTC, January 2005.
9. "Scalable Network Technologies, Inc., QualNet simulator," available at http://www. qualnet.com

Improving Multicast Stability in Mobile Multicast Scheme Using Motion Prediction*

Qian Wu[1], Jianping Wu[1], and Mingwei Xu[2]

[1] Department of Computer Science and Technology, Tsinghua University,
Beijing 100084, China
{wuqian, jianping}@cernet.edu.cn
[2] Department of Computer Science and Technology, Tsinghua University,
Beijing 100084, China
xmw@csnet1.cs.tsinghua.edu.cn

Abstract. Stability is an important issue in multicast, especially in mobile environment where joining and leaving behaviors occur much more frequently. In this paper, we propose a scheme to improve the multicast stability by the use of motion prediction. The mobile node (MN) predicts the staying time before entering the new network, if the time is long enough, it will ask the new network to join the multicast tree as usual. Otherwise, the new network should create a tunnel to the multicast agent of MN to receive multicast packets. Considering that networks usually have different power range, the staying time is not predicted directly, and the Average Staying Time is used instead. The prediction algorithm is effective but practical which requires little calculation time and memory size. The simulation results show that the proposed scheme can improve the stability of multicast tree remarkably while bring much smaller cost.

1 Introduction

Mobile communication is playing an increasing role in our lives. Nowadays the most widely used mobility pattern is host mobility, often called *terminal mobility*, and this kind of mobile users always expect similar applications to static ones. Providing multicast in mobile networks is inspirited because IP multicast can provide many attractive applications, such as video conferencing, multiplayer games, etc, and bring the merit of efficient multi-destinations delivery which can save network bandwidth and release the burden of replications from the source.

In mobile environment, multicast must deal with not only dynamic group membership but also dynamic member location. But current multicast protocols are developed implicitly for static members and do not consider the extra requirements to support mobile nodes. Every time a member changes its location, keeping track of it and reconstructing the multicast tree will cause extreme overhead, while leaving the multicast tree unchanged will result in inefficient

---

* This work is supported by the Natural Science Foundation of China (No 60373010 and No 90604024), and National 973 Project Fund of China (No 2003CB314801).

A. Helmy et al. (Eds.): MMNS 2006, LNCS 4267, pp. 235–246, 2006.

sometimes incorrect delivery path. Thus, providing multicast support for mobile nodes is challenging and faces with many new problems [2, 3], such as routing inefficiency, multicast group and tree maintenance overhead, handoff latency, packets loss and multicast tree stability.

Stability is an important issue in multicast, especially in mobile environment where joining and leaving behaviors occur much more frequently. Former study [6] shows that when the number of group members is not bigger than the 1/3 of network size, the multicast tree would be an unstable one. Wu *et al* [7] find that the stability problem is more serious in mobile environment and it is mainly dominated by three elements, namely the ratio of the number of mobile nodes and network size, mobility model and the mobile multicast scheme.

In this paper we introduce a technique to improve the stability of multicast tree by the use of motion prediction, and apply it on the FHMM (Mobile Multicast with Fast Handoff and Hierarchical Architecture) [1] mobile multicast scheme. FHMM scheme has many advantages such as low packet loss rate, high multicast packet delivery efficiency and little multicast maintenance overhead. But it can be still improved in the aspect of multicast tree stability. The new scheme proposed in this paper is called P-FHMM. It uses a practical motion prediction algorithm to predict the staying time of MN in the new network. If the expected staying time is long enough, this network will be asked to join the multicast group as same as in FHMM. Otherwise, this network should only create a tunnel to the Multicast Agent (MA) of MN and use this tunnel to receive multicast packets. Simulation results show that, the stability of multicast tree can be obviously improved while its expense - the efficiency - will only decrease a little. While building tunnel will cause some additional handoff latency, the simulation results show that the packet loss percentage of P-FHMM scheme is still very small and within the acceptable range.

In section 2 we briefly introduce some mobile multicast schemes and the stability issue of multicast tree. Section 3 describes P-FHMM scheme and the motion prediction algorithm in detail. Section 4 presents our experiments using OMNet++ [11] and results analysis. Finally, we conclude this paper.

2   Background and Related Works

2.1   Mobile Multicast Schemes

Mobile IP&IPv6 [5,4] provide two basic approaches to support mobile multicast, i.e., bi-directional tunneling (BT) and remote subscription (RS). Most of the other proposed mobile multicast solutions are based on them [2, 3].

In BT approach, MN subscribes to multicast group through its home agent (HA), and uses the bi-directional tunnel between them to receive multicast packets. In this manner, there is no need to update multicast group state and multicast tree after MN's handoff, so there is no multicast maintenance overhead. But the multicast delivery path of MN is far from optimal, and routing inefficiency and bandwidth wasting become the main drawbacks.

In RS approach, MN always re-subscribes to the multicast group when it changes the attachment to a new access network. It maintains most of the merits of multicast mechanism especially the high packet delivery efficiency, and people pay more attentions to it and propose many mobile multicast solutions based on it. But the main problem of RS and RS-based approaches is the overhead of maintaining the multicast delivery tree because joining and leaving behaviors occur much more frequently in mobile networks.

FHMM scheme [1] is based on RS approach and solves its main drawbacks. It introduces two kinds of agents: *Multicast Agent* (MA) and *Domain Multicast Agent* (DMA). MA is the extended Access Router (AR) responsible for providing multicast service to MNs in the access network. DMA is a multicast enabled router which manages multicast within its domain. Through the use of these two kinds of agents, FHMM achieves a hierarchical mobile multicast architecture which can shield the local movement from outside, efficiently reduce the change of inter-domain multicast delivery tree and decrease the multicast protocol overhead related to handoff. In addition, FHMM introduces a fast multicast handoff technology through which the AR of expected foreign network (called as *EAR*) can be notified to join the dedicated multicast group in advance, which helps MN to receive multicast packets more quickly after the handoff. Then both the handoff latency and the packet loss rate can be reduced.

But FHMM scheme does not solve the stability problem of intra-domain multicast tree. When the number of group member is not very big and the move speed of MN is fast, it will happen that the MN moves out of the range of the newly visited network just after this network having joined the multicast tree, or sometimes the MN may move to other networks before this network finishing joining the multicast tree. In this situation, the intra-domain tree will be caused to change frequently. What's more, due to the overlapping of networks, its fast multicast handoff mechanism may cause several networks to join the multicast group at the same time which would result in more serious stability problem.

2.2   Stability of Multicast Tree

Mieghem *et al* [6] studies how the number of links in a multicast tree changes after one multicast member leaves the group. For a multicast group with m different members distributed in the graph containing N nodes, the average number of link changes is denoted by $E[\Delta_N(m)]$, and when the expression $E[\Delta_N(m)] \leq 1$ is satisfied, the author claims the multicast tree would be a stable one.

Wu *et al* [7] studies the stability of IP multicast tree in mobile environment. In its expanded definition of stability, the link change of the multicast tree can be caused by three kinds of multicast group membership updates, namely join, leave and leave-join event. The author calls the value $E[\Delta_N(m)]$ as *Stability Factor*. The simulation results show that stability problem is more serious in mobile environment and the Stability Factor is mainly dominated by three elements. The first is the ratio of m/N which reflects the density of mobile nodes in the network and the probability of MN moving to a network which already has MNs in its range. The second is the mobility model which effects how these mobile nodes are distributed in the network, and the aggregative degree of them. The third is the mobile multicast scheme which influences the way of updating multicast tree when a position change event happens. Although the effect of speed of MN and the power range of AR on the Stability Factor remains slight, both of them will cause the multicast tree to update more frequently.

3   Proposed Scheme

3.1   Overview

FHMM can reduce the change of inter-domain multicast tree, but the multicast tree inside the domain remains unimproved. And due to the overlapping of networks, its fast multicast handoff mechanism may cause more than one networks to join the multicast group at the same time which would result in more serious stability problem. Especially when the number of group members is not very big and the move speed of mobile nodes is fast, it will cause more transitory or sometimes unuseful joining actions, and then aggravate the stability problem.

In this paper, we propose P-FHMM scheme to improve the multicast stability of FHMM by the use of motion prediction. Every time the MN detects the connection to a new access network, it predicts the staying time in this network. If the expected staying time is long enough, this network will be asked to join the multicast group as same as in FHMM. Otherwise, this network should only create a tunnel to the MA of MN and use this tunnel to receive multicast packets. By the use of motion prediction, the short-term and unnecessary joining group action can be eliminated and the stability can be improved. What's more, P-FHMM only makes few modifications on FHMM and achieves visible performance.

P-FHMM can use any motion prediction algorithm to predict staying time. Here we use the Adaptive Response Rate Single Exponential Smoothing (ARRSES) [8] method. It is a simple but effective algorithm with little calculation and memory requirements. It is practical in mobile environment since the capability of mobile device is always limited than static one. What's more, in view that network may have different power range, we do not predict staying time directly. We use ARRSES algorithm to predict the average staying time in the new network instead, and then the expected staying time can be worked out.

3.2   Operation Details

Modifications of FHMM

As mentioned in [1], the operation of FHMM include two main procedures, the first is multicast group joining procedure and the second is handoff procedure. The multicast group joining procedure of P-FHMM is the same as FHMM, while the fast multicast handoff procedure has some modifications.

The first modification is adding a "T" flag in the multicast group option in fast multicast handoff procedure. If the "T" flag is set, it means the prediction result made by MN is using tunnel. Otherwise, the result is joining multicast group as usual. Through the exchange of messages carrying multicast group option, the EAR in the new network can be informed with the prediction result.

The second modification is adding a new message named FNABack (Fast Neighbor Advertisement Acknowledge). In FHMM scheme, as soon as the MN is handoff to a new network, it sends a FNA (Fast Neighbor Advertisement) message to announce its attachment. We add this new FNABack message to acknowledge FNA message, and thus MN can know the ultimate handoff decision of EAR in this new network and then change its own handoff decision and modify its MA information correspondingly. It is used to ensure efficiency. For example, if the "T" flag is set but the EAR has already been on the multicast tree, there is no need to create a tunnel transferring the same multicast packets.

The last modification is adding the power range information of EAR in PrRtAdv (Proxy Router Advertisement) message. In FHMM scheme, the PrRtAdv message is used to tell the MN the address information of EAR. After this modification, the MN can also know the power range of EAR and this information is needed in our prediction algorithm.

Handoff procedure

This section introduces the multicast handoff procedure of P-FHMM in detail, especially the difference with FHMM scheme.

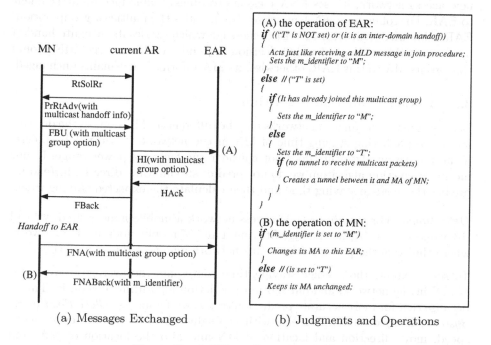

(a) Messages Exchanged   (b) Judgments and Operations

Fig. 1. Handoff Procedure of P-FHMM

Multicast handoff in P-FHMM is still fast and efficient. Fig 1 shows the hand-off . As we can see, the main operations of multicast handoff procedure change a little. Fig 1(a) illustrates the messages exchanged in handoff procedure. They are the same as FHMM scheme except the new FNABack message used in the final stage. Fig 1(b) shows the detailed judgments and operations of MN and EAR.

When MN discovers the availability of a new access network by some link-specific methods while still connected to its current network, it starts fast multicast handoff procedure. By the information of EAR's power range gained in PrRtAdv message, MN predicts the staying time in the new network. If this expected staying time does not reach the pre-assigned value, the prediction result is using tunnel, and the "T" flag is set. Otherwise the "T" flag should be cleared. Then through the use of FBU (Fast Binding Update) and HI (Handover Initiate) messages, MN can transfer multicast group option carrying "T" flag information to EAR which would motivate EAR to determine the ultimate handoff decision. EAR will make the ultimate handoff decision to joining multicast tree in three situations: the "T" flag in multicast group option is not set which means the expected staying time is long enough, or EAR itself has already joined this multicast tree which means there is no need to create a tunnel anymore, or the handoff is an inter-domain one and EAR should join the multicast tree in order to keep the hierarchical architecture. Otherwise the ultimate handoff decision is set to using tunnel and EAR should create a tunnel to the MA of MN.

When the MN is disconnected from the previous network and moves to the new access network, it uses FNA message to quickly announce its attachment to EAR. For robustness, FNA message also includes the multicast group option. EAR acknowledges it with FNABack message which carries its ultimate handoff decision. If this ultimate handoff decision is joining multicast tree, MN should changes its MA to this EAR, otherwise, its MA information remains unchanged.

3.3   Motion Prediction Algorithm

Because the power range of networks may be different and it is one of the decisive factors of predicting staying time of MN in the network, it will not be precise if the prediction algorithm does not consider the factor of power range. In our motion prediction algorithm we do not predict staying time directly, instead we predict the average staying time and then calculate the expected staying time.

Definition 1. For the *ith* ($i \geq 1$) access network a mobile node visited, we call the value $\eta_i = T_i/R_i$ as Average Staying Time of mobile node in this network, where the R_i is the power range of this network and T_i is the staying time.

To some extent, the average staying time represents the average move speed of MN in the network, and our motion prediction algorithm runs on it. This is because that for more accurate prediction result, we should use *Globe Positioning System* (GPS) or something else with the same effect to get the exact move speed, move direction and location of MN and also the location of MA, and then we can predict the staying time more precisely. But the GPS is not always available and the capability of mobile device is usually limited than static one, the burdensome calculation of prediction is not always acceptable. What's more, the mobility of MN has the property of randomness, even if MN knows all the precise information and has enough calculation capability, the exact value can not always be predicted.

Facing with the fact that a mobile user usually travels with a destination in mind, and the change of mobile node's velocity within a short time is limited due to physical restrictions, we predict the future average move speed, i.e. average staying time, of MN correlating to the previous and current ones.

The prediction method we use is the *Adaptive Response Rate Single Exponential Smoothing* (ARRSES) algorithm introduced by Trigg and Leach [8]. It is a very popular and effective forecasting method with the merit of easy to use, very little calculation and memory requirements, the ability of automatic adjusting the predicting parameter to forecasting error. It is a variation of Exponential Smoothing method and can continually adjust the *Smoothing Parameter* α_i.

The expected average staying time ($\tilde{\eta}_{n+1}$) is calculated as follows:

$$\tilde{\eta}_{n+1} = \alpha_n \times \eta_n + (1 - \alpha_n) \times \tilde{\eta}_n \tag{1}$$

where the $\tilde{\eta}_{n+1}$ and $\tilde{\eta}_n$ are the prediction value of $n+1$ and n network, η_n is the real value of n network, and α_n is smoothing parameter to predict the value in $n+1$ network and $0 \leq \alpha_n \leq 1$. α_n is calculated by following steps:

$$e_n = \eta_n - \tilde{\eta}_n \tag{2}$$

$$E_n = \beta e_n + (1 - \beta) E_{n-1} \tag{3}$$

$$M_n = \beta |e_n| + (1 - \beta) M_{n-1} \tag{4}$$

$$\alpha_n = |E_n / M_n| \tag{5}$$

where the parameter β is usually set at 0.1 or 0.2.

Then the expected staying time can be worked out by $\tilde{T}_{n+1} = \tilde{\eta}_{n+1} \times R_{n+1}$, and if $\tilde{T}_{n+1}$ is not smaller than the threshold value T_{th} , the handoff decision is joining multicast tree and the "T" flag should be cleared, otherwise the handoff decision is using tunnel and the "T" flag should be set.

4   Performance Evaluation

4.1   Network Model and Methodology

The simulation is built on OMNET++ [11], a discrete event simulator. We compare the stability of P-FHMM with FHMM in terms of stability factor and the multicast tree change frequency. We also investigate the main costs of improving stability - the packet delivery efficiency and the packet lost rate.

The topology in our simulation is a 10×10 mesh network in which each node acts as a multicast router of local network and also an MA for MN. There are 4 DMA routers in the topology, and each DMA is responsible for 25 (5×5) MAs. The power range is 71m, and the distance between two nearby MAs is 100m.

For simplicity, there is one multicast group with one data source. The source generates multicast packet with the length of 300 bytes at intervals of 20 ms to simulate the multicast conference application.

In our simulation, multicast group members are all mobile nodes, and the update of multicast tree is absolutely caused by the position change events of mobile members. The number of mobile nodes varies from 5 to 30. Originally, mobile nodes are randomly located at the mesh network.

For all the performances investigated, we check them both in Random Waypoint [9] (RPW for short) and Gauss-Markov [10](GM for short) mobility model. In RPW model, the MN at first chooses a random destination in the simulation area and a speed. It then travels toward this new destination. Upon arrival, it pauses for a random period before starting the process again. In our simulation we set the pause time to 0 to evaluate the performance in most random condition. The speed is uniformly chosen between 0 and 30 m/s.

While in GM mobility model, MN's new speed and direction are correlated to its formal speed and direction and their mean values through the simulation. This model uses a tuning parameter, α, where $0 \leq \alpha \leq 1$, to vary the randomness. The smaller the α, the greater the randomness is. We choose the value of α to 0.8 to simulate the move with high historical character, and the mean value of speed is fixes to 30m/s which reflects the MN with high speed.

We run each simulation for 500 seconds.

4.2   Simulation Results

Fig 2 to 8 are the simulation results. To investigate the P-FHMM scheme, we examine three different staying time thresholds, namely 3, 5 and 10 seconds.

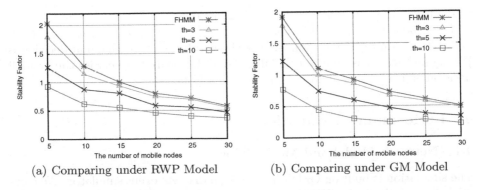

(a) Comparing under RWP Model   (b) Comparing under GM Model

Fig. 2. Comparison of Stability Factor

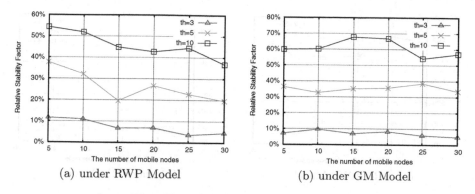

Fig. 3. Relatively Reduced Value of Stability Factor

Fig 2 compares the stability factor. Fig 3 is the relatively reduced value of stability factor of P-FHMM which reflects the improvement on FHMM scheme. From the four pictures in these two figures we can see: 1) through the use of motion prediction, the stability factor of mobile multicast scheme can be improved; 2) the longer the staying time threshold of joining multicast tree, the smaller the stability factor which means the more stable of the multicast tree; 3) when the staying time threshold is about 5 seconds, the stability can be improved by about 30% and the stability factor is nearly all below 1, the stable regime defined in [6] and [7].

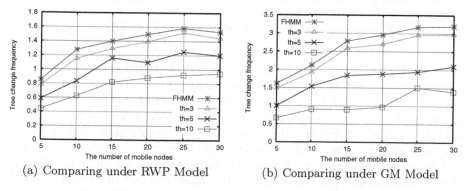

Fig. 4. Comparison of Multicast Tree Change Frequency

Fig 4 and 5 illustrate the simulation results of multicast tree change frequency. Fig 4 compares the absolute value and fig 5 is the relatively reduced value. From the figures we can conclude that through the using of motion prediction, the frequency of modifying multicast tree can be reduced remarkably. And the longer the staying time threshold of joining multicast tree will result in less modification frequency and better improvement of stability.

What's more, we find that fig 5 is very similar to fig 3. Stability factor and multicast tree change frequency are two main elements of investigating multicast

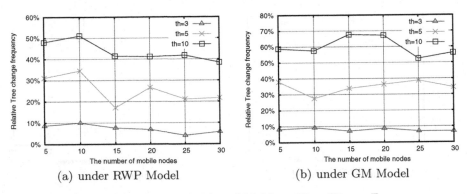

(a) under RWP Model   (b) under GM Model

Fig. 5. Relatively Reduced Value of Multicast Tree Change Frequency

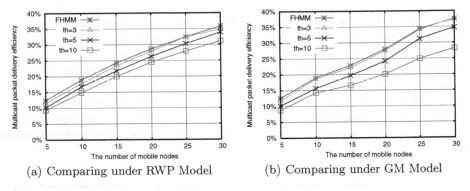

(a) Comparing under RWP Model   (b) Comparing under GM Model

Fig. 6. Comparison of Multicast Packet Delivery Efficiency

stability. The similarity of these two figures reflects that P-FHMM scheme can equally improve both of these two stability elements, which verifies its effectivity.

For the use of tunnel, the main cost of improving stability is the packet delivery efficiency. Fig 6 and 7 compare the multicast packet delivery efficiency of P-FHMM and FHMM. Fig 6 is the comparison of absolute value and fig 7 is the relatively reduced value. From these two figures we can have some conclusions. Firstly, improving stability will certainly reduce the packet delivery efficiency. This is because the pure multicast is the most efficient mechanism and the use of tunnel will decrease a part of multicast to unicast. Secondly, the longer the staying time threshold is, the more inefficient the mobile multicast scheme will be. This is because that longer threshold will result in more tunnels to be used. Thirdly, and also the most important, as shown in fig 7 the decreasing of efficiency is much smaller (nearly the half of) than the improvement of stability illustrated in fig 3 and fig 5. Let's look at line of threshold 5 seconds for example. The improvement of stability of the proposed P-FHMM scheme in fig 3 and fig 5 is around 30%, while the cost, namely decreased efficiency shown in fig 7, is just around 15% . This result means P-FHMM scheme uses little cost to gain more significant improvement, and further verifies its effectivity.

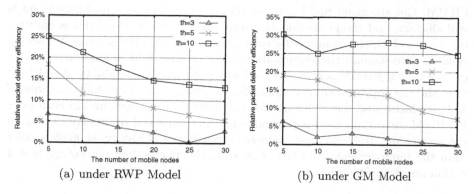

(a) under RWP Model   (b) under GM Model

Fig. 7. Relatively Reduced Value of Multicast Packet Delivery Efficiency

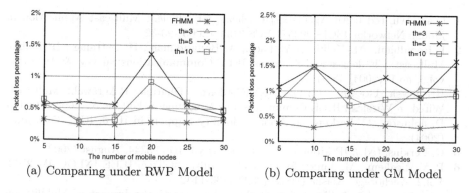

(a) Comparing under RWP Model   (b) Comparing under GM Model

Fig. 8. Comparison of Multicast Packet Lost Percentage

Fig 8 is the comparison of multicast packet lost percentage. Obviously the FHMM scheme achieves the best performance. For the need of building the tunnel, P-FHMM introduces new handoff latency and this would cause more packets to be lost. But the packet lost percentage keeps relatively uninfluenced with the staying time threshold, and although more packets have been lost, the packet lost percentage of P-FHMM is still around 0.5% under RPW model and 1% under GM model. These results are still acceptable for multicast applications.

5   Conclusion and Future Work

In this paper, we propose a scheme (P-FHMM) to improve the multicast stability of FHMM by the use of motion prediction. In the multicast handoff procedure of P-FHMM scheme, MN predicts its staying time in the new network at first. If the expected staying time is long enough, it will ask the new network to join the multicast tree as usual. Otherwise, the new network will only be asked to create a tunnel to the multicast agent of MN and use this tunnel to receive multicast packets. To achieve this, the P-FHMM scheme makes some modifications on

FHMM. But the main multicast handoff procedure remains unchanged and is still efficient, and the prediction algorithm introduced by P-FHMM is effective but practical which requires little calculation time and memory size. We set up simulation to evaluate the performance. From the abundant simulation results we can see that P-FHMM scheme is effective in improving the stability of multicast tree, and the improvement is remarkable. We investigate the stability in terms of stability factor and the multicast tree modification frequency, and in both sides P-FHMM can obtain good performance and the improved results are similar. In the mean time, the main cost, namely decreasing of efficiency, is much smaller than the improvement. And the other cost, the packet lost percentage is still within an acceptable range.

References

1. Qian, Wu. (ed.): An Agent-Based Scheme for Efficient Multicast Application in Mobile Networks. ISCIS 2005, LNCS 3733. (2005) 23-32
2. I, Romdhani., M, Kellil., H-Y, Lach., A, Bouabdallah., H, Bettahar.: IP Mobile Multicast: Challenges and Solutions. IEEE Communications Surveys & Tutorials, vol.6, no.1. (2004) 18-41
3. Hrishikesh, Gossain., Carlos, De.M.Cordeiro., Dharma, P.Agrawal.: Multicast: Wired to Wireless. IEEE Communications Magazine, vol.40, no.6. (2002) 116-123
4. D, Johnson., C, Perkins., J, Arkko.: Mobility Support in IPv6. RFC 3775, June (2004)
5. C, Perkins. (ed.): IP Mobility Support for IPv4. RFC 3344, August (2002)
6. P, V.Mieghem., M, Janic.: Stability of a Multicast Tree. IEEE INFOCOM 2002, Vol.2. Anchorage, Alaska (USA) (2002)1133-1142
7. Qian Wu. (ed.): A Simulation Study to Investigate the Impact of Mobility on Stability of IP Multicast Tree. MSN 2005, LNCS 3794. (2005) 836-845
8. D. W. Trigg., D. H. Leach.: Exponential Smoothing with an Adaptive Response Rate, Operational Research Quarterly, vol. 18. (1967) 53-59
9. D, Johnson., D, Maltz.: Dynamic source routing in ad hoc wireless networks. In T. Imelinsky and H. Korth, editors, Mobile Computing. (1996) 153-181.
10. B, Liang., Z, Haas.: Predictive distance-based mobility management for PCS networks. Proceedings of the Joint Conference of the IEEE Computer and Communications Societies (INFOCOM). (1999) 1377-1384
11. Omnet++ Community Site. http://www.omnetpp.org

End-to-End QoS Issues of MPEG4-FGS Video Streaming Traffic Delivery in an IP/UMTS Network

Thomas Pliakas[1], George Kormentzas[1], and Charalabos Skianis[1,2]

[1] Universiy of Aegean,
Department of Information and Communication Systems Engineering,
GR-83200, Karlovassi, Greece
{tpliakas, gkorm}@aegean.gr
[2] National Centre for Scientific Research 'Demokritos',
Institute of Informatics & Telecommunications
skianis@iit.demokritos.gr

Abstract. The paper addresses the end-to-end QoS problem of MPEG4-FGS video streaming traffic delivery over a heterogeneous IP/UMTS network. It proposes and validates an architecture that explores the joint use of packet prioritization and scalable video coding together with the appropriate mapping of UMTS traffic classes to the DiffServ traffic classes. A set of simulation scenarios, involving eight different video sequences, demonstrates the quality gains of both scalable video coding and prioritized packetization.

Keywords: DiffServ, End-to-End QoS, MPEG4-FGS, Packet Prioritization, UMTS.

1 Introduction

Fixed and wireless/mobile network operators face a twin challenge: to create and deliver attractive IP-based multimedia services quickly in response to fast-changing business and customer demands; and to evolve their current underlying networking infrastructures to an architecture that can deliver such services in a highly adaptable and guaranteed end-to-end Quality of Service (QoS) way both from network and application perspectives.

Simultaneously, on the customer side, for the next few years at least, there will be a wide variety of mobile/wireless access technologies supporting IP connectivity. These technologies include: mobile communication networks, such as GPRS [1] and UMTS [2], the family of broadband radio access networks, like IEEE 802.11 [3] and HIPERLAN [4], and wireless broadcasting technologies, like digital video broadcasting (DVB—satellite and terrestrial [5]).

IP technology seems to be able to resolve the interworking amongst the diverse fixed core and wireless/mobile access technologies at the network level. In this all-IP network, the end-to-end QoS provision concerning the network perspective could be established through the appropriate mapping amongst the QoS traffic classes/services supported by the contributing underlying networking technologies. Building on this context, this work concerns a DiffServ-aware IP core network and a UMTS access

A. Helmy et al. (Eds.): MMNS 2006, LNCS 4267, pp. 247–255, 2006.

network and examines end-to-end QoS issues regarding MPEG4-FGS video streaming traffic delivery over such a network.

The Differentiated Services (DiffServ) model proposed by IETF support (based on the DSCP field of the IP header) two different services, the Expedited Forwarding (EF) that offers low packet loss and low delay/jitter and the Assured Forwarding (AF), which provides QoS guarantees better than the best-effort service. Differences amongst AF services imply that a higher QoS AF class will give a better performance (faster deliver, lower loss probability) than a lower AF class [6].

The QoS provision in Universal Mobile Telecommunications System (UMTS) is achieved through the concept of "bearers". A bearer is a service providing a particular QoS level between two defined points invoking the appropriate schemes for either the creation of QoS guaranteed circuits, or the enforcement of special QoS treatments for specific packets. The selection of bearers with the appropriate characteristics constitutes the basis for the UMTS QoS provision. Each UMTS bearer is characterized by a number of quality and performance factors. The most important factor is the bearer's Traffic Class; four traffic classes have been defined in the scope of the UMTS framework (i.e., Conversational, Streaming, Interactive and Background). The appropriate mapping of UMTS traffic classes to the aforementioned DiffServ service classes could offer a vehicle for the end-to-end QoS provision over a heterogeneous DiffServ/UMTS network. In our work, we employ and evaluate the three different mapping approaches presented in [7]-[9] respectively.

The Fine Grain Scalability (FGS) [10] feature of MPEG4 is a promising scalable video solution to address the problem of guaranteed end-to-end QoS provision concerning the application perspective. According to MPEG4-FGS, the Base Layer (BL) provides the basic video quality to meet the minimum user bandwidth, while the Enhancement Layer (EL) can be truncated to meet the heterogeneous network characteristics, such as available bandwidth, packet loss, and delay/jitter [11].

To address the end-to-end QoS problem of MPEG4-FGS video streaming traffic delivery over a heterogeneous IP/UMTS network, the paper proposes and validates through a number of NS2-based simulation scenarios an architecture that explores the joint use of packet prioritization and scalable video coding together with the appropriate mapping of UMTS traffic classes to the DiffServ traffic classes.

The rest of the paper is organized as follows. In Section 2, the proposed video coding and prioritization framework for providing QoS guarantees for MPEG4-FGS video streaming traffic delivery over a heterogeneous IP/UMTS network is presented. In Section 3, we demonstrate how video-streaming applications can benefit from the use of the proposed architecure. Finally, Section 4 draws the conclusions of this work.

2   Overview of the Proposed Architecture

Our architecture integrates the concepts of MPEG4-FGS video streaming, prioritized packetization based on content and DiffServ/UMTS classes coupling. The proposed architecture is depicted in Figure 1. It consists of three key components: (1) MPEG4-FGS scalable video encoding, (2) simple prioritized packetization according to the type of content (I, P, B frame type), and (3) DiffServ/UMTS classes coupling in order to achieve QoS continuity of MPEG4-FGS video streaming traffic delivery over DiffServ and UMTS network domains. Each one of these components is discussed in detail in the following subsections.

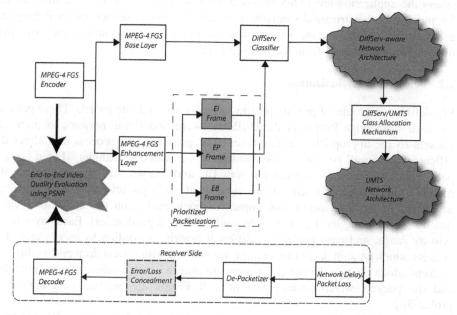

Fig. 1. Overview of proposed architecture

2.1 MPEG4-FGS Scalable Video Coding

MPEG4-FGS scalable video coding constitutes a new video coding technology that increases the flexibility of video streaming. Similar to the conventional scalable encoding, the video is encoded into a BL and one or more ELs. For MPEG4-FGS, the EL can be efficient truncated in order to adapt transmission rate according to underlying network conditions. This feature can be used by the video servers to adapt the streamed video to the available bandwidth in real-time (without requiring any computationally demanding re-encoding). In addition, the fine granularity property can be exploited by the intermediate network nodes (including base stations, in case of wireless networks) in order to adapt the video stream to the currently available downstream bandwidth.

In contrast to conventional scalable methods, the complete reception of the EL for successful decoding is not required [13]. The received part can be decoded, increasing the overall video quality according to the rate-distortion curve of the EL as it described [14]. The overall video quality can be also improved from the error concealment method that is used. In our architecture, when a frame is lost, the decoder inserts a successfully previous decoded frame in the place of each lost frame. A packet is also considered as lost, if the delay of packet is more than the time of the play-out buffer. (For the experiments discussed in the next Section III, this time is set to 1sec).

In order to measure the improvements in video quality by employing MPEG4-FGS, we adopt the Peak Signal-to-Noise Ratio (PSNR) and Structural SIMilarity (SSIM) [12] metrics. PSNR is one of the most commonly used objective metric to assess the application-level QoS of video transmissions and SSIM is a novel metric for measuring the structural similarity between two image sequences, exploiting the general principle that the main function of the human visual system is the extraction of structural information from the viewing field.

2.2 Prioritized Packetization

We define two groups of priority policies, one for BL and one for EL. These policies are used from Edge Router of the DiffServ-aware underlying network to mark the packets to the appropriate traffic classes. The packetization process can affect the efficiency as well as the error resiliency of video streaming. Fixed length packetization scheme is adopted for both BL and EL streams as proposed by the MPEG4 specification. Based on the content of each packet, we assign priorities according to the anticipated loss impact of each packet on the end-to-end video quality (considering the loss impact to itself and to dependencies). Each layer has a priority range, and each packet has different priority according to its payload. The packets which contain data of an I Frame are marked with lowest drop probability, the packets which contain data of a P Frame are marked with medium drop probability and the packets which contain data of a B Frame are marked with high drop probability.

Note that MPEG4-FGS specification assumes guaranteed delivery to BL and best-effort one to EL. In our framework, we use EF for transmitting BL and AF with different priorities for the EL based on the frame type. With assigned priorities, the packets are sent to underlying network and receive different forwarding treatments. Table 1 depicts the relation between the type of the EL content and the corresponding DiffServ classes. The first digit of the AF class indicates forwarding priority and the second indicates the packet drop precedence.

Table 1. DiffServ classes allocation for EL

Frame Type	DiffServ Classes
I Frame	AF11
P Frame	AF12
B Frame	AF13

2.3 DiffServ/UMTS Classes Coupling

The proposed MPEG4-FGS video streaming traffic delivery framework adopts three different coupling of DiffServ/UMTS classes approaches depicted in Table 2. Note that the actual QoS that can be obtained heavily depends on the traffic engineering for both UMTS and DiffServ networks.

Table 2. DiffServ/UMTS classes coupling

DiffServ Classes	UMTS Traffic Classes (Setting I) [7]	UMTS Traffic Classes (Setting II) [8]	UMTS Traffic Classes (Setting III) [9]
EF	Streaming	Conversational	Conversational
AF11	Interactive 1	Streaming	Streaming
AF12	Interactive 2	Streaming	Streaming
AF13	Interactive 3	Streaming	Interactive
BE	Background	Background	Background

3 Framework Evaluation

This section evaluates the performance of the proposed architectural framework through a set of experimental cases. A NS2- based simulation environment with the appropriate EURANE package extensions for simulating a UMTS network is adopted. We study the performance of our framework by enabling or disabling scalable video coding and/or by enabling or disabling prioritized transmission. The quality gains of scalable video coding in comparison with non-scalable video coding and the quality gains of prioritized transmission in comparison with non-prioritized transmission applying three different DiffServ/UMTS traffic classes mapping approaches are discussed in detail.

Fig. 2 depicts our simulation setup, which includes a DiffServ-aware autonomous system of a single 512Kbps wired link and a single UMTS cell of 1Mbps with the following rate allocation for the supported traffic classes: 200Kbps for the Conversional class, 300Kbps for the Streaming class, 200kbps for the Interactive 1 class, 100kbps for both Interactive 2 and 3 classes, and 200Kbps for the Background class. For the DiffServ-aware network the buffer management is considered to be WRED. The qualitative remarks being the outcome of our experiments can be also applied over more complex heterogeneous IP/UMTS infrastructures.

Several YUY QCIF (176x144) video sequences consisting of 300 to 2000 frames are used as video sources. A number of background flows are also transmitted in the simulated network in order to fill in the respective DiffServ/UMTS class capacity in the link. The background traffic is increased from 210Kbps to 540Kbps leading the system in congestion.

The validation of the quality gains offered by the proposed framework concerns four simulation cases consisting in a number of experiments referring to eight different source video sequences transmissions over an all-IP network consisting of a DiffServ-aware IP core network and a UMTS access one.

The first simulation case refers to a single layer MPEG4 stream transmission. The video frames are sent every 33ms for 30fps video. For this simulation scenario, we use EF for transmitting I frames and AF12 and AF13 for transmitting P and B frames respectively. The mapping of DiffServ classes to the UMTS ones is performed through Table 2.

The second simulation case concerns a scalable MPEG4 stream transmission consisting in two layers. The BL packets are encoded using the MPEG4-FGS codec with MPEG2 TM5 rate control at 128kbps and the EL ones are encoded at 256kbps. For this case, we have direct application of Tables 1 and 2.

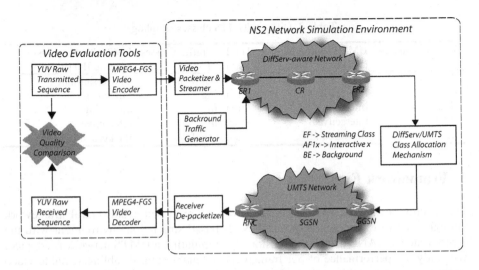

Fig. 2. Simulation Setup

The third simulation case concerns a scalable MPEG4 stream transmission consisting in one BL and two ELs, i.e., EL1 and EL2. The encoding of BL packets remains at 128kbps as in the second simulation case, while the encoding of packets of both ELs is at 128kbps. For this simulation scenario, we use EF for transmitting BL, AF11 for transmitting EL1, and Best Effort (BE) for transmitting EL2. The mapping of DiffServ classes to the UMTS ones is performed through Table 2.

The fourth simulation case adopts the setup of the third case, while it applies the prioritized packetization scheme of the second case to the packets of the first EL, i.e., for this simulation scenario, we use EF for transmitting BL, Table 1 for transmitting EL1, and Best Effort (BE) for transmitting EL2.

Tables 3 to 5 depict the simulation results in terms of PSNR and SSIM video quality metrics for eight different YUV video sequences for all simulation cases (1 to 4) for the three settings (I to III) concerning Diffserv/UMTS classes coupling. For Setting I, each configuration case increases the video quality and the gain increment that offers each case is around 2db in terms of PSNR. For Setting II, the Cases 3 and 4 produce the same results.

Table 3. Quality Results for all Simulation Cases for Diffserv/UMTS classes coupling of Setting I

Video Sequences	Number of Frames	Case 1		Case 2		Case 3		Case 4	
		PSNR	SSIM	PSNR	SSIM	PSNR	SSIM	PSNR	SSIM
Bridge-Close	2001	25.455	0.0673	27.025	0.0772	29.565	0.0815	31.026	0.0896
Highway	2000	28.321	0.0761	30.658	0.0874	31.875	0.0937	33.451	0.0986
Grandma	871	28.365	0.0761	29.982	0.0832	31.453	0.0905	32.821	0.0949
Claire	494	27.981	0.0731	30.025	0.0896	31.751	0.0936	32.973	0.0978
Salesman	444	28.456	0.0762	31.563	0.0912	32.961	0.0957	34.361	0.0985
Foreman	400	29.012	0.0816	31.454	0.0905	33.568	0.0982	34.816	0.0993
Carphone	382	25.565	0.0675	28.234	0.0796	31.028	0.0896	32.564	0.0942
Container	300	24.545	0.0684	27.194	0.0784	29.729	0.0829	31.581	0.0912

Table 4. Quality Results for all Simulation Cases for Diffserv/UMTS classes coupling of Setting II

Video Sequences	Number of Frames	Case 1		Case 2		Case 3		Case 4	
		PSNR	SSIM	PSNR	SSIM	PSNR	SSIM	PSNR	SSIM
Bridge-Close	2001	32.123	0.0938	32.783	0.0942	29.213	0.0817	29.218	0.0817
Highway	2000	34.342	0.0987	34.632	0.0989	31.321	0.0908	31.341	0.0908
Grandma	871	34.943	0.0991	34.232	0.0984	31.763	0.0919	31.768	0.0919
Claire	494	33.231	0.0979	33.683	0.0977	32.497	0.0926	31.591	0.0927
Salesman	444	35.039	0.0996	35.913	0,0999	31.938	0.0937	31.942	0.0937
Foreman	400	35.725	0.0998	35.281	0.0997	32.321	0.0944	31.327	0.0943
Carphone	382	33.184	0.0983	33.432	0.0987	31.293	0.0915	31.284	0.0913
Container	300	32.718	0.0948	32.782	0.0948	29.123	0.0817	29.128	0.0817

Table 5. Quality Results for all Simulation Cases for Diffserv/UMTS classes coupling of Setting III

Video Sequences	Number of Frames	Case 1		Case 2		Case 3		Case 4	
		PSNR	SSIM	PSNR	SSIM	PSNR	SSIM	PSNR	SSIM
Bridge-Close	2001	32.118	0.0936	33.562	0.0968	29.218	0.0818	29.217	0.0817
Highway	2000	34.212	0.0985	34.432	0.0987	31.319	0.0909	31.314	0.0907
Grandma	871	34.679	0.0988	34.782	0.0989	31.764	0.0917	31.763	0.0917
Claire	494	33.235	0.0979	33.783	0.0978	31.497	0.0925	31.489	0.0923
Salesman	444	34.671	0.0988	34.732	0.0990	31.942	0.0937	31.936	0.0937
Foreman	400	34.983	0.0995	35. 243	0.0997	32.316	0.0329	32.297	0.0328
Carphone	382	32.928	0.0953	33.421	0.0973	31.292	0.0982	31.286	0.0979
Container	300	32.594	0.0941	33.783	0.0978	29.432	0.0821	29.425	0.0821

For the Highway video sequence, we measure the packet/frame losses for I, P, and B frames for the four simulation cases for the three settings (I to III) concerning Diffserv/UMTS classes coupling. For Cases 3 and 4 the depicted measurements concern EL1. The results presented in Tables 6-8 are in accordance with the ones depicted in Tables 3-5. For Setting I, each case improves the previous one and Case 4 offers the best video quality gain as it experiences the lower packet/frame losses. For Settings II and III, Case 2 offers the best video quality.

Table 6. Packet/Frame losses for the Highway video sequence for Diffserv/UMTS classes coupling of Setting I

Frame Type	Case 1 Frame Loss	Case 1 Packet Loss	Case 2 - EL Frame Loss	Case 2 - EL Packet Loss	Case 3 - EL1 Frame Loss	Case 3 - EL1 Packet Loss	Case 4 - EL1 Frame Loss	Case 4 - EL1 Packet Loss
I	0.1%	3,4%	0.1%	3,1%	0.1%	2,1%	0.1%	0.1%
P	11.4%	12.6%	11.1%	11.9%	10.7%	11.8%	5.7%	6,3%
B	47,3%	47.7%	43,6%	43.9%	42,6%	42.8%	23.9%	27.8%

Table 7. Packet/Frame losses for the Highway video sequence for Diffserv/UMTS classes coupling of Setting II

Frame Type	Case 1 Frame Loss	Case 1 Packet Loss	Case 2 - EL Frame Loss	Case 2 - EL Packet Loss	Case 3 - EL1 Frame Loss	Case 3 - EL1 Packet Loss	Case 4 - EL1 Frame Loss	Case 4 - EL1 Packet Loss
I	0.1%	3.2%	0.1%	2.4%	0.1%	2.7%	0.1%	3.1%
P	6.3%	7.5%	5.7%	6.2%	5.6%	6.8%	5.5%	6.1%
B	19.7%	12.7%	16.7%	9.8%	15.6%	8.7%	15.4%	8.9%

Table 8. Packet/Frame losses for the Highway video sequence for Diffserv/UMTS classes coupling of Setting III

Frame Type	Case 1 Frame Loss	Case 1 Packet Loss	Case 2 - EL Frame Loss	Case 2 - EL Packet Loss	Case 3 - EL1 Frame Loss	Case 3 - EL1 Packet Loss	Case 4 - EL1 Fra me Loss	Case 4 - EL1 Packet Loss
I	0.0%	0.0%	0.1%	1.8%	0.1%	1.2%	0.1%	1.7%
P	5.2%	7,8%	6.8%	11.3%	6.4%	7.2%	6.7%	7.1%
B	22.7%	23,8%	21.9%	23.5%	15.3%	17.1%	24.3%	26.8%

As an overall remark of the above results, we could note that Case 4 of Setting I could offer almost the same video quality as Case 2 of Settings II and III, without however employing conversational class.

4 Conclusions

Nowadays, continuous media applications over heterogeneous all-IP networks, such as video streaming and videoconferencing, are become very popular. Several approaches have been proposed in order to address the end-to-end QoS both from network perspective, like DiffServ and UMTS QoS traffic classes, and from application perspective, like scalable video coding and packetized prioritization mechanisms. The paper addresses the end-to-end QoS problem of MPEG4-FGS video streaming traffic delivery over a heterogeneous IP/UMTS network. It proposes and validates through a number of NS2-based simulation scenarios a framework that explores the joint use of packet prioritization and scalable video coding together with the appropriate mapping of UMTS traffic classes to the DiffServ traffic classes.

Acknowledgment

The work reported in this paper is carried out within the project "Study and Development of Interactive Broadband Services based on DVB-T/DVB-H Technologies" in the context of framework 2.2 of 'Pythagoras II – Research Group Support of the University of the Aegean' jointly funded by the European Union and the Hellenic Ministry of Education.

References

1. 3GPP, General Packet Radio Service (GPRS); Service Description, Tech. Spec. 3GPP TS 23.060. v3.12.0, June 2002; http://www.3gpp.org
2. Richardson KW. UMTS overview. *Journal of Electronics and Communication Engineering* 2000; 12(3): 93–100.
3. IEEE Std 802.11, 1999 Edition (ISO/IEC 8802-11: 1999).
4. Khun-Jush J, Schramm P, Malmgren G, Torsner J. 'HiperLAN2: broadband wireless communications at 5 GHz'. *IEEE Communication Magazine* 2002; 40(6): 130–136.
5. ETSI, Digital Video Broadcasting (DVB); Framing structure, channel coding and modulation for digital terrestrial television, EN300-744, January 2001.
6. IETF Web site: http://www.ietf.org, DiffServ and IntServ Working Groups.
7. S. Maniatis, E. Nikolouzou and I. Venieris, "QoS Issues in the Converged 3G Wireless and Wired Networks", *IEEE Communication Magazine*, Aug. 2002.
8. H. Wang, D. Prasad, O. Teyeb and H-P. Schwefel, "Performance Enhancements of UMTS networks using end-to-end QoS provisioning", IWS 2005/WPMC'05, Aalborg, Denmark. Sept. 2005
9. R. Chakravorty, I. Pratt and J. Crowcroft, "A Framework for Dynamic SLA-based QoS Control for UMTS", *IEEE Communication Magazine*, Oct. 2003
10. W. Li, "Overview of fine granular scalability in mpeg-4 video standard", *IEEE Transaction on Circuits and Systems for Video Technology*, vol. 11, no. 3, pp. 301–317, 2001.
11. H. M. Radha, M. van de Schaar, and Y. Chen, "The MPEG-4 fine grained scalable video coding method for multimedia streaming over IP", *IEEE Transactions on Multimedia*, vol. 3, no. 1, pp. 53–68, 2001.
12. Zhou Wang, Ligang Lu, and Alan C. Bovik, "Video Quality Assessment Based on Structural Distortion Measurement", *Signal Processing: Image Communication*, special issue on "Objective Video Quality Metrics", vol. 19, no. 2, pp. 121-132, Feb. 2004.
13. P. de Cuetos, M. Reisslein, and K. W. Ross, "Evaluating the streaming of fgs-encoded video with rate-distortion traces", Institute Eurecom Technical Report, Tech Report RR-03-078, 2003
14. P. Seeling, P. de Cuetos, and M. Reisslein, "Fine granularity scalable video: Implications for streaming and a trace based evaluation methodology", IEEE Communication Magazine, vol. 43, no. 4, pp. 138–142, 2005.

Author Index

Lecture Notes in Computer Science

For information about Vols. 1–4172

please contact your bookseller or Springer

Vol. 4270: H. Zha, Z. Pan, H. Thwaites, A.C. Addison, M. Forte (Eds.), Interactive Technologies and Sociotechnical Systems. XVI, 547 pages. 2006.

Vol. 4267: A. Helmy, B. Jennings, L. Murphy, T. Pfeifer (Eds.), Autonomic Management of Mobile Multimedia Services. XIII, 257 pages. 2006.

Vol. 4265: N. Lavrač, L. Todorovski, K.P. Jantke (Eds.), Discovery Science. XIV, 384 pages. 2006. (Sublibrary LNAI).

Vol. 4264: J.L. Balcázar, P.M. Long, F. Stephan (Eds.), Algorithmic Learning Theory. XIII, 393 pages. 2006. (Sublibrary LNAI).

Vol. 4254: T. Grust, H. Höpfner, A. Illarramendi, S. Jablonski, M. Mesiti, S. Müller, P.-L. Patranjan, K.-U. Sattler, M. Spiliopoulou (Eds.), Current Trends in Database Technology – EDBT 2006. XXXI, 932 pages. 2006.

Vol. 4253: B. Gabrys, R.J. Howlett, L.C. Jain (Eds.), Knowledge-Based Intelligent Information and Engineering Systems, Part III. XXXII, 1301 pages. 2006. (Sublibrary LNAI).

Vol. 4252: B. Gabrys, R.J. Howlett, L.C. Jain (Eds.), Knowledge-Based Intelligent Information and Engineering Systems, Part II. XXXIII, 1335 pages. 2006. (Sublibrary LNAI).

Vol. 4251: B. Gabrys, R.J. Howlett, L.C. Jain (Eds.), Knowledge-Based Intelligent Information and Engineering Systems, Part I. LXVI, 1297 pages. 2006. (Sublibrary LNAI).

Vol. 4249: L. Goubin, M. Matsui (Eds.), Cryptographic Hardware and Embedded Systems - CHES 2006. XII, 462 pages. 2006.

Vol. 4248: S. Staab, V. Svátek (Eds.), Engineering Knowledge in the Age of the Semantic Web. XIV, 400 pages. 2006. (Sublibrary LNAI).

Vol. 4247: T.-D. Wang, X. Li, S.-H. Chen, X. Wang, H. Abbass, H. Iba, G. Chen, X. Yao (Eds.), Simulated Evolution and Learning. XXI, 940 pages. 2006.

Vol. 4243: T. Yakhno, E.J. Neuhold (Eds.), Advances in Information Systems. XIII, 420 pages. 2006.

Vol. 4241: R.R. Beichel, M. Sonka (Eds.), Computer Vision Approaches to Medical Image Analysis. XI, 262 pages. 2006.

Vol. 4239: H.Y. Youn, M. Kim, H. Morikawa (Eds.), Ubiquitous Computing Systems. XVI, 548 pages. 2006.

Vol. 4238: Y.-T. Kim, M. Takano (Eds.), Management of Convergence Networks and Services. XVIII, 605 pages. 2006.

Vol. 4236: L. Breveglieri, I. Koren, D. Naccache, J.-P. Seifert (Eds.), Fault Diagnosis and Tolerance in Cryptography. XIII, 253 pages. 2006.

Vol. 4234: I. King, J. Wang, L. Chan, D. Wang (Eds.), Neural Information Processing, Part III. XXII, 1227 pages. 2006.

Vol. 4233: I. King, J. Wang, L. Chan, D. Wang (Eds.), Neural Information Processing, Part II. XXII, 1203 pages. 2006.

Vol. 4232: I. King, J. Wang, L. Chan, D. Wang (Eds.), Neural Information Processing, Part I. XLVI, 1153 pages. 2006.

Vol. 4229: E. Najm, J.F. Pradat-Peyre, V.V. Donzeau-Gouge (Eds.), Formal Techniques for Networked and Distributed Systems - FORTE 2006. X, 486 pages. 2006.

Vol. 4228: D.E. Lightfoot, C.A. Szyperski (Eds.), Modular Programming Languages. X, 415 pages. 2006.

Vol. 4227: W. Nejdl, K. Tochtermann (Eds.), Innovative Approaches for Learning and Knowledge Sharing. XVII, 721 pages. 2006.

Vol. 4225: J.F. Martínez-Trinidad, J.A. Carrasco Ochoa, J. Kittler (Eds.), Progress in Pattern Recognition, Image Analysis and Applications. XIX, 995 pages. 2006.

Vol. 4224: E. Corchado, H. Yin, V. Botti, C. Fyfe (Eds.), Intelligent Data Engineering and Automated Learning – IDEAL 2006. XXVII, 1447 pages. 2006.

Vol. 4223: L. Wang, L. Jiao, G. Shi, X. Li, J. Liu (Eds.), Fuzzy Systems and Knowledge Discovery. XXVIII, 1335 pages. 2006. (Sublibrary LNAI).

Vol. 4222: L. Jiao, L. Wang, X. Gao, J. Liu, F. Wu (Eds.), Advances in Natural Computation, Part II. XLII, 998 pages. 2006.

Vol. 4221: L. Jiao, L. Wang, X. Gao, J. Liu, F. Wu (Eds.), Advances in Natural Computation, Part I. XLI, 992 pages. 2006.

Vol. 4219: D. Zamboni, C. Kruegel (Eds.), Recent Advances in Intrusion Detection. XII, 331 pages. 2006.

Vol. 4218: S. Graf, W. Zhang (Eds.), Automated Technology for Verification and Analysis. XIV, 540 pages. 2006.

Vol. 4217: P. Cuenca, L. Orozco-Barbosa (Eds.), Personal Wireless Communications. XV, 532 pages. 2006.

Vol. 4216: M.R. Berthold, R. Glen, I. Fischer (Eds.), Computational Life Sciences II. XIII, 269 pages. 2006. (Sublibrary LNBI).

Vol. 4215: D.W. Embley, A. Olivé, S. Ram (Eds.), Conceptual Modeling - ER 2006. XVI, 590 pages. 2006.

Vol. 4213: J. Fürnkranz, T. Scheffer, M. Spiliopoulou (Eds.), Knowledge Discovery in Databases: PKDD 2006. XXII, 660 pages. 2006. (Sublibrary LNAI).

Vol. 4212: J. Fürnkranz, T. Scheffer, M. Spiliopoulou (Eds.), Machine Learning: ECML 2006. XXIII, 851 pages. 2006. (Sublibrary LNAI).

Vol. 4211: P. Vogt, Y. Sugita, E. Tuci, C. Nehaniv (Eds.), Symbol Grounding and Beyond. VIII, 237 pages. 2006. (Sublibrary LNAI).

Vol. 4210: C. Priami (Ed.), Computational Methods in Systems Biology. X, 323 pages. 2006. (Sublibrary LNBI).

Vol. 4209: F. Crestani, P. Ferragina, M. Sanderson (Eds.), String Processing and Information Retrieval. XIV, 367 pages. 2006.

Vol. 4208: M. Gerndt, D. Kranzlmüller (Eds.), High Performance Computing and Communications. XXII, 938 pages. 2006.

Vol. 4207: Z. Ésik (Ed.), Computer Science Logic. XII, 627 pages. 2006.

Vol. 4206: P. Dourish, A. Friday (Eds.), UbiComp 2006: Ubiquitous Computing. XIX, 526 pages. 2006.

Vol. 4205: G. Bourque, N. El-Mabrouk (Eds.), Comparative Genomics. X, 231 pages. 2006. (Sublibrary LNBI).

Vol. 4204: F. Benhamou (Ed.), Principles and Practice of Constraint Programming - CP 2006. XVIII, 774 pages. 2006.

Vol. 4203: F. Esposito, Z.W. Raś, D. Malerba, G. Semeraro (Eds.), Foundations of Intelligent Systems. XVIII, 767 pages. 2006. (Sublibrary LNAI).

Vol. 4202: E. Asarin, P. Bouyer (Eds.), Formal Modeling and Analysis of Timed Systems. XI, 369 pages. 2006.

Vol. 4201: Y. Sakakibara, S. Kobayashi, K. Sato, T. Nishino, E. Tomita (Eds.), Grammatical Inference: Algorithms and Applications. XII, 359 pages. 2006. (Sublibrary LNAI).

Vol. 4200: I.F.C. Smith (Ed.), Intelligent Computing in Engineering and Architecture. XIII, 692 pages. 2006. (Sublibrary LNAI).

Vol. 4199: O. Nierstrasz, J. Whittle, D. Harel, G. Reggio (Eds.), Model Driven Engineering Languages and Systems. XVI, 798 pages. 2006.

Vol. 4198: O. Nasraoui, O. Zaiane, M. Spiliopoulou, B. Mobasher, B. Masand, P. Yu (Eds.), Advances in Web Mining and Web Usage Analysis. IX, 177 pages. 2006. (Sublibrary LNAI).

Vol. 4197: M. Raubal, H.J. Miller, A.U. Frank, M.F. Goodchild (Eds.), Geographic, Information Science. XIII, 419 pages. 2006.

Vol. 4196: K. Fischer, I.J. Timm, E. André, N. Zhong (Eds.), Multiagent System Technologies. X, 185 pages. 2006. (Sublibrary LNAI).

Vol. 4195: D. Gaiti, G. Pujolle, E. Al-Shaer, K. Calvert, S. Dobson, G. Leduc, O. Martikainen (Eds.), Autonomic Networking. IX, 316 pages. 2006.

Vol. 4194: V.G. Ganzha, E.W. Mayr, E.V. Vorozhtsov (Eds.), Computer Algebra in Scientific Computing. XI, 313 pages. 2006.

Vol. 4193: T.P. Runarsson, H.-G. Beyer, E. Burke, J.J. Merelo-Guervós, L.D. Whitley, X. Yao (Eds.), Parallel Problem Solving from Nature - PPSN IX. XIX, 1061 pages. 2006.

Vol. 4192: B. Mohr, J.L. Träff, J. Worringen, J. Dongarra (Eds.), Recent Advances in Parallel Virtual Machine and Message Passing Interface. XVI, 414 pages. 2006.

Vol. 4191: R. Larsen, M. Nielsen, J. Sporring (Eds.), Medical Image Computing and Computer-Assisted Intervention – MICCAI 2006, Part II. XXXVIII, 981 pages. 2006.

Vol. 4190: R. Larsen, M. Nielsen, J. Sporring (Eds.), Medical Image Computing and Computer-Assisted Intervention – MICCAI 2006, Part I. XXXVVIII, 949 pages. 2006.

Vol. 4189: D. Gollmann, J. Meier, A. Sabelfeld (Eds.), Computer Security – ESORICS 2006. XI, 548 pages. 2006.

Vol. 4188: P. Sojka, I. Kopeček, K. Pala (Eds.), Text, Speech and Dialogue. XV, 721 pages. 2006. (Sublibrary LNAI).

Vol. 4187: J.J. Alferes, J. Bailey, W. May, U. Schwertel (Eds.), Principles and Practice of Semantic Web Reasoning. XI, 277 pages. 2006.

Vol. 4186: C. Jesshope, C. Egan (Eds.), Advances in Computer Systems Architecture. XIV, 605 pages. 2006.

Vol. 4185: R. Mizoguchi, Z. Shi, F. Giunchiglia (Eds.), The Semantic Web – ASWC 2006. XX, 778 pages. 2006.

Vol. 4184: M. Bravetti, M. Núñez, G. Zavattaro (Eds.), Web Services and Formal Methods. X, 289 pages. 2006.

Vol. 4183: J. Euzenat, J. Domingue (Eds.), Artificial Intelligence: Methodology, Systems, and Applications. XIII, 291 pages. 2006. (Sublibrary LNAI).

Vol. 4182: H.T. Ng, M.-K. Leong, M.-Y. Kan, D. Ji (Eds.), Information Retrieval Technology. XVI, 684 pages. 2006.

Vol. 4180: M. Kohlhase, OMDoc – An Open Markup Format for Mathematical Documents [version 1.2]. XIX, 428 pages. 2006. (Sublibrary LNAI).

Vol. 4179: J. Blanc-Talon, W. Philips, D. Popescu, P. Scheunders (Eds.), Advanced Concepts for Intelligent Vision Systems. XXIV, 1224 pages. 2006.

Vol. 4178: A. Corradini, H. Ehrig, U. Montanari, L. Ribeiro, G. Rozenberg (Eds.), Graph Transformations. XII, 473 pages. 2006.

Vol. 4177: R. Marín, E. Onaindía, A. Bugarín, J. Santos (Eds.), Current Topics in Artificial Intelligence. XV, 482 pages. 2006. (Sublibrary LNAI).

Vol. 4176: S.K. Katsikas, J. Lopez, M. Backes, S. Gritzalis, B. Preneel (Eds.), Information Security. XIV, 548 pages. 2006.

Vol. 4175: P. Bücher, B.M.E. Moret (Eds.), Algorithms in Bioinformatics. XII, 402 pages. 2006. (Sublibrary LNBI).

Vol. 4174: K. Franke, K.-R. Müller, B. Nickolay, R. Schäfer (Eds.), Pattern Recognition. XX, 773 pages. 2006.

Vol. 4173: S. El Yacoubi, B. Chopard, S. Bandini (Eds.), Cellular Automata. XV, 734 pages. 2006.